The Case for Perfection

Johann A. R. Roduit

The Case for Perfection
Ethics in the Age of Human Enhancement

Bibliographic Information published by the Deutsche Nationalbibliothek
The Deutsche Nationalbibliothek lists this publication in the Deutsche Nationalbibliografie; detailed bibliographic data is available in the internet at http://dnb.d-nb.de.

Cover photo: © Olivier Lovey
'Adam n°3457b'

Library of Congress Cataloging-in-Publication Data
Names: Roduit, Johann A. R., 1982- author.
Title: The case for perfection : ethics in the age of human enhancement / Johann A.R. Roduit.
Description: 1 [edition]. | New York : Peter Lang, 2016. | Includes bibliographical references and index.
Identifiers: LCCN 2015046780 | ISBN 9783631671504
Subjects: LCSH: Perfection. | Biotechnology--Moral and ethical aspects. | Bioethics-Philosophy.
Classification: LCC BD233 .R63 2016 | DDC 128-dc23 LC record available at http://lccn.loc.gov/2015046780

ISBN 978-3-631-67150-4 (Print)
E-ISBN 978-3-653-06516-9 (E-Book)
DOI 10.3726/978-3-653-06516-9

© Peter Lang GmbH
Internationaler Verlag der Wissenschaften
Frankfurt am Main 2016
All rights reserved.
Peter Lang Edition is an Imprint of Peter Lang GmbH.

Peter Lang – Frankfurt am Main · Bern · Bruxelles · New York · Oxford · Warszawa · Wien

All parts of this publication are protected by copyright. Any utilisation outside the strict limits of the copyright law, without the permission of the publisher, is forbidden and liable to prosecution. This applies in particular to reproductions, translations, microfilming, and storage and processing in electronic retrieval systems.

This publication has been peer reviewed.

www.peterlang.com

To Jessica

Perfection is attained by slow degrees; it requires the hand of time.
— Voltaire

When you aim for perfection, you discover it's a moving target.
— George Fisher

Table of Contents

Preface by Nicholas Agar 13

Acknowledgements 15
    Contributions 16

Summary 19

Clarifying

Introduction 23
    Background and evidence of the problem 23
        Definitions of human enhancement 23
        The ethical debate 24
        The anthropological arguments 26
        Working definition of human perfection 26
    Research question and thesis 28
        The concept of perfection is unavoidable 28
        Which conception of perfection should we use? 31
    Outline 33
    Implications 35
    Scope and limitations 36

Chapter 1: Living up to the ideal human 39
    Summary 39
    Introduction 39
    The conundrum of defining human enhancement 39
        Beyond: the 'beyond therapy' approach 40
        More: the 'quantitative' approach 41
        Better: the 'qualitative' approach 42

Limitations of ethical tools used to evaluate human enhancement 42
Safety 43
Justice 44
Autonomy 45
Anthropological arguments 46
Living up to the ideal human: evaluating human enhancement 47
Advantages of this approach 51
Conclusion 52

Criticizing

Chapter 2: Human enhancement and perfection 57
Summary 57
Introduction 57
Bioconservatives and perfection 58
Skepticism regarding perfection 59
The paradoxical argument against perfection 60
Bioliberals and perfection 63
Can bioliberals do without perfection? 64
Conclusion 66

Conceptualizing

Chapter 3: Evaluating human enhancements: the importance of ideals 71
Summary 71
Ideal and non-ideal views about human enhancement 71
Human enhancement as improvement 72
Two ways of looking at improvement: backward and forward 73
Ideal and non-ideal approaches to evaluate human enhancement 74

Limitations of a non-ideal approach 76
    Lack of an objective and shortsightedness 76
    The appearance of neutrality 78
Defending an ideal approach 79
    Insufficient 79
    Unnecessary 80
    Pluralism and sufficiency 81
    Inflexible 82
Conclusion 83

Constructing

Chapter 4: Rejecting problematic conceptions of perfection 87

Summary 87
Introduction 87
Content of Perfection 88
Sources of Perfection 91
    Problems with a subjective approach 92
    Problems with an objective approach 93
Function of Perfection 94
From human perfection to perfectionist notions 95
Perfectionist assumptions in the debate 96
    Michael Sandel 96
    The President's Council on Bioethics and Leon Kass 98
    Nick Bostrom 100
    Nicholas Agar 102

Chapter 5: Defending a particular view of perfection 105

Summary 105
Introduction 105

11

Defining capabilities 105
Reasons to look at the capabilities approach 106
    Make it explicit 106
    Being human: objective type-perfection 107
    Ideal theory 107
    End-state thinking 108
    Room for pluralism 108
    Holistic approach of the capabilities approach 109
    Both constraining and guiding 110
    Conclusions 110

Concluding

Conclusion 115

Bibliography 119

Index 127

Preface by Nicholas Agar

In his work, Johann Roduit makes a fascinating case for a view of human enhancement centered on the concept of perfection, in which he presents perfection as an objective ideal, one informed by the study of actual human characteristics. Roduit offers an illuminating commentary on the apparent aversion to the concept expressed by many participants in the philosophical debate about enhancement. He suggests that the concept of perfection is nonetheless an unacknowledged assumption to the views of many philosophers who explicitly disavow any interest in it.

I must confess that I initially approached Roduit's project with some skepticism. I first encountered the concept of perfection within the philosophical debate about enhancement in political philosopher Michael Sandel's article "The Case Against Perfection," published in the April 2004 issue of U.S. magazine *The Atlantic Monthly*. As a relatively unknown philosopher myself, I was flattered to be quoted in the piece – admittedly, as one of the "bad" guys. In Sandel's critique of enhancement, I was cast in the role of the philosophical naïf who thinks that genetic enhancement can extend procreative liberty rather than being used to restrict it. I proposed that parents could think about selecting for their children's genes in a similar way to how they currently go about selecting their children's educational influences (see Sandel, 2007). When it comes to educating children in a liberal society, parents do not have an entirely free hand: Minimum standards must be achieved and certain eccentric choices are justly prohibited. I continue to hold this view. The first thing that struck me was how wrongheaded the title of Sandel's article seemed, given the fact that no philosophically responsible advocate of enhancement actually conceives of perfection as their aim. The concept of perfection seems to me an unwanted and unnecessary transplant from theological debates: Perfection may be available to an omnipotent deity (though religious philosophers suggest that this is not entirely unproblematic, even in this context), but no combination of pedagogical, genetic, or cybernetic techniques or technologies could ever result in human perfection. The parallel between genetic and educational choices suggests to me that the concept of improvement is a far better fit in the enhancement debate. For instance, parents who give their children after-school tutoring in mathematics seek to improve their children's mathematical abilities. They are not aiming at achieving mathematical perfection – whatever that might even be. This also applies to those who seek to tweak the genes that influence mathematical ability. Originally, I had imagined the title *The Case Against Perfection* was forced on Sandel by an editor as a provocative, attention-grabbing headline, but this assumption proved less likely with the publication of his 2007 book under the same title.

Roduit responds to my skepticism, and his work on the concept of perfection is a valuable contribution to the debate. In this study, Roduit goes far beyond using the concept as a not-entirely-accurate label for a position he seeks to reject. Instead, he makes positive and creative use of the concept by exploring issues that are left obscured by other approaches that use enhancement a synonym for improvement. Prospective parents using genetic technologies to select certain characteristics in their children tend to have more immediate and short-range concerns. Some of these are adequately captured by a conception of enhancement as improvement. When parents alter genes that influence intelligence, they are content with the likelihood that their child will be more intelligent than she would otherwise have been. But when we ask questions about enhancement specifically, we also have longer-range concerns, and this is where Roduit's perfectionist account proves so valuable. An enhancement is not just an improvement; it is "also an improvement leading somewhere." Roduit's concept of enhancement thus enables us to ask important questions about "what type of human are we hoping to become."

You might wonder whether a need a new concept of enhancement is necessary. Indeed, philosophers tend to excel at inventing new concepts, which helps to explain our unfortunate reputation in the wider academe as unhelpful hair-splitters. Scholars active in other disciplines, who prefer dealing with more concrete issues, are told that they need to learn a whole lot of new words to describe phenomena that they credit themselves as already understanding well enough. But we need more words – not fewer – to describe the diverse and unexpected ways in which genetic, cybernetic, and other technologies might alter us. Roduit's perfectionist account of enhancement is a very welcome conceptual addition to the debate and deserves to be widely read. It should generate a richer, if not perfect, debate on human enhancement.

– Nicholas Agar

Acknowledgements

I am grateful first and foremost to Professor Nikola Biller-Andorno, who has welcomed me to the Institute for Biomedical Ethics and History of Medicine (IBME) at the University of Zurich. It has been my privilege to work in such a stimulating environment, which allowed me the freedom to explore new ideas.

This work could not have been accomplished without the help of both my co-supervisors. I am grateful to Dr. Jan-Christoph Heilinger for always challenging my work and giving me the tools to further explore and develop my own views, and to Dr. Holger Baumann for the countless hours spent discussing ideas and refining my work. Without their help, this project would have most likely never have seen the light of day. My collaboration with both Dr. Heilinger and Dr. Baumann on various papers eventually culminated in this doctoral work.

I am also very grateful to Prof. Julian Savulescu, who hosted me for a research stay at the Uheiro Center of Practical Ethics at the University of Oxford. While at Oxford, I had the privilege to present my work at various events, where I was provided with valuable feedback from participants, particularly from Dr. Tom Douglas, my supervisor while in Oxford.

I would like to offer a very special thank you to the colleagues and friends I met at the University of Zurich, especially to Prof. Effy Vayena, Dr. Roberto Andorno, Dr. Carina Fourie, Dr. Regula Ott, Dr. Jürg Streuli, Dr. Zümrüt Alpinar, and Corine Dorey, who, in one way or another, greatly contributed to this research.

I am also grateful to the organizers and attendees of the Hastings-Brocher Summer Academy on Human Enhancement at the Brocher Foundation in Geneva, which took place during the earlier stages of my project in June 2012. I have subsequently had the opportunity to meet some of them again at various conferences around the world. Dr. Cynthia Forlini and Pieter Bonte provided valuable feedback on my work. I also owe a particular thank you to Dr. Vincent Menuz and Dr. Alexandre Erler, with whom I launched the think tank NeoHumanitas in an effort to bring public awareness of the use of emerging and future technologies outside of the walls of the academe.

I also really appreciate the input I received over the years from Michael Buttrey, Nathan Phillips, Jake Benjamins, Jenny Williams, and Rebecca Koenig, who patiently reviewed segments of my work.

Parts of this study have previously been presented at different conferences, congresses, workshops, and seminars around the world. I would like to thank the attendees of these past events for their helpful comments, encouragement and critiques, especially the participants of the *25th European Conference on Philosophy of Medicine and Health Care* in at the University of Zurich (August 2011); the *11th IAB World Congress of Bioethics* in Rotterdam (June 2012); the conference on *Transforming the Human: Enhancement, Emerging Technologies, and Social Challenges* in Dubrovnik, Croatia (September 2012); the Autumn Seminar of the *Swiss Society of Biomedical Ethics* in Bigorio, Switzerland (November 2012); the Applied Ethics Graduate Discussion Group led by Julian Savulescu at the Oxford Uehiro Center; the conference *The Posthuman: Differences, Embodiments, Performativity* at the Universita'di Roma Tre in Rome (September 2013); the *Autumn Seminar* of the Swiss Society of Biomedical Ethics in Bigorio, Switzerland (November 2013); the *12th IAB World Congress of Bioethics* in Mexico City (June 2014); and finally, participants of the *II International Workshop in Practical Ethics: Bioethics and Human Enhancement* at the University of Granada in Spain (June 2014).

While writing the research proposal, my work was generously funded by the University Research Programme for Ethics (UFSP) of the University of Zurich and the IBME. Subsequent funding for my dissertation was provided by the Swiss National Science Foundation (SNSF, Project 141419). The research for the first article was co-funded by the Käthe Zingg-Schwichtenberg Fund from the Swiss Academy of Medical Sciences (SAMS). Parts of the dissertation were written during a research stay at the University of Oxford, thanks to a research fellowship awarded by the SNSF.

Finally, I am grateful to my family and friends for their ongoing encouragement, and last, but not least, to my wife Jessica for her unconditional support. This work is dedicated to her.

Contributions

I produced all the original material of this thesis, which consists of four articles, as well as the introduction and conclusion. Drafts of the first article were critically revised by Vincent Menuz and Holger Baumann. The article titled "Human enhancement: Living up to the ideal human" was published as a peer-reviewed book chapter in *Global Issues and Ethical Considerations in Human Enhancement Technologies*, (Ed. Steve J. Thompson, IGI Global, www.igi-global.com, 2014, pp. 54–66). This chapter is reproduced here with permission from the publisher.

Drafts of the second and third manuscripts were also critically revised by Holger Baumann and Jan-Christoph Heilinger. The second article has been reproduced here from the journal article "Human enhancement and perfection" (*Journal of Medical Ethics*, 39(10), 2013: 647–50), with permission from the BMJ Publishing Group Ltd. The third article was published as "Evaluating human enhancements: The importance of ideals" (*Monash Bioethics Review*, 32(3), 2015: 205–216). It is reprinted here with permission from Springer Science+Business Media.

Finally, the draft of the final article also benefited from valuable comments from Holger Baumann and Jan-Christoph Heilinger. Parts of the manuscript were presented at the *12th World Congress of Bioethics* in Mexico City, Mexico (June 26, 2014) and at the *II International Workshop in Practical Ethics: Bioethics and Human Enhancement* at the University of Granada, Spain (June 16, 2014). The manuscript "Ideas of Perfection and the Ethics of Human Enhancement" was published in the journal *Bioethics*, 29, 2015: 622–630. In this book, the last article has been divided into two more developed chapters (Chapters 4 and 5) for the sake of clarity and to properly delineate the arguments. Parts of the last article are reprinted here by permission of the publisher.

Summary

This doctoral dissertation critically examines what role – if any – the notion of perfection should play in the debate of the ethics of human enhancement. After distinguishing between the concept of perfection and different philosophical conceptions of perfection, I first argue that the concept of perfection is unavoidable in the context of this debate, as the notion of human enhancement itself is inextricably interconnected with the notion of human perfection. This leads to the next issue of what particular conception of perfection should be used in the debate, and in what ways. After rejecting various problematic possibilities, I argue that the concept of human perfection should fulfill certain criteria: It should be an objective ideal, yet not holding a fixed view of what a human being ought to be. This position allows for pluralism and political deliberation, while at the same time provides guidance, not only restrictions, to the real-world practice of human enhancement. I show here how philosopher Martha Nussbaum's 'capabilities approach' fulfills these criteria, making it a tool that can be used to morally assess human enhancement in addition to other bioethical standards, such as safety, justice, and autonomy.

This research is located, therefore, in a balanced position between the so-called bioconservative stance, which uses some particular conceptions of perfection to argue against human enhancement, and the so-called bioliberal view, which also, at times, uses some specific conceptions of perfection (or rather, as I will further illustrate, some particular perfectionist notions of what it means to live a good life), in order to argue in favor of human enhancement. This thesis demonstrates that a particular conception of human perfection – based on a harmonized combination of Nussbaum's central capabilities – is in some cases *for* human enhancement, and surprisingly enough, *against* it in other circumstances. Ultimately, this theoretical approach to the conception of human perfection helps us to distinguish between true human enhancement and so-called dis-enhancement.

Clarifying

Introduction

Background and evidence of the problem

Even though much has been written and debated on the ethics of human enhancement, there is still little agreement concerning what exactly human enhancement is, or how to morally evaluate it.

Definitions of human enhancement

Three leading definitions have emerged in this debate. First, human enhancement has been defined in relation to *therapy*. In this view, human enhancement is a medical or a technological intervention that does not attempt to cure, but rather improves an individual's capabilities in a particular area above and beyond what is considered to be normal human functioning. This particular intervention is regarded to be outside the scope or the aims of traditional medicine, which seeks to cure by restoring a sick individual to normal functioning.

Second, human enhancement is also occasionally understood as adding a particular characteristic to an individual by giving them a capacity or capability they did not have before. In this *quantitative* understanding, human enhancement does not necessarily make a person better because it merely adds something to an individual's current state, but does not qualitatively assess the effects of this addition on the person.

Finally, the third definition of human enhancement refers to *qualitative* change. Here, human enhancement is considered to be an improvement for the better. This third understanding overlaps with the first definition, as both address the qualitative aspect of change. The difference, however, is that the first definition contrasts enhancement with therapy, whereas the second does not. As such, the third definition implies that the treatment of a disease can also be understood as a form of enhancement. These definitions of human enhancement can be applied to both a particular human capacity (e.g. cognition) or to the individual as a whole.

I will later defend a qualitative understanding of human enhancement that takes into consideration the person as a whole (Chapter 1). In the meantime, one can already question the idea of improvement by asking questions such as "Better according to what? Or whom?"[1] Two approaches – which have not often been identified in the literature – may help answer these questions. On the one hand, the notion of "better" can be considered an improvement over a former state, or what I call a *backward-looking view*. Here, two states or conditions are compared with one another in order to decide which one is better than the other. On the other hand, an improvement can be considered as such by referencing it to a certain ideal in what I refer to as the *forward-looking view*. This view adds a reference point that can be used as a standard by which to compare the two other states of being. Here, I refer to this reference point as perfection, the end of the trajectory along which human enhancement can aim.

The ethical debate

Despite a lack of a consensus regarding a conclusive definition of human enhancement, the ethical debate has nevertheless developed quite rapidly. Different normative tools have been used to assess whether enhancement should be morally permissible, prohibited, encouraged or even mandatory, resulting in the emergence of three distinct positions regarding the morality of human enhancement. In the first school of thought, *Bioconservatives* regard human enhancement as morally problematic due to the fact that it involves too many risks and carries the possibility of unintended consequences that could potentially have a negative impact on human dignity or human nature. In the second position, *Bioliberals* argue that while some forms of human enhancement may be morally problematic, this is not necessarily the case in every instance of enhancement. Bioliberals thus argue that individuals should be free to enhance so long as enhancements are used fairly and safely. Finally, *Transhumanists* seek to use human enhancement as a tool to improve the human condition beyond what is considered normal, even if this requires becoming posthuman – something other than human. One transhumanist goal, for example, is to eradicate disease and aging, which would result in human lifespan being extended far beyond what is biologically possible or seen as 'human' today.

---

1   Kass rightly asks, "Enhancement is, even as a term, highly problematic. Does it mean 'more' or 'better,' and if 'better,' by what standard?" (Kass, 2003).

While these three categories can be helpful in establishing a broad overview of the debate, the reality is much more complex, and it is helpful to consider different moral positions regarding human enhancement as a spectrum of views between the two extremes of the bioconservative and the transhumanist positions. On the one side of the spectrum, it is argued that any enhancement is morally problematic and should be banned. At the other end of the spectrum, others would argue that some forms of enhancement should be mandatory in order to improve the lives of human beings, even if this means becoming posthuman. And between these two extremes, there are a variety of competing, nuanced positions.[2]

In this ethical debate, the bioethical notions of justice, safety, and autonomy (Chapter 1) have often been deployed to argue both in favor and against enhancement. However, these normative tools alone are not sufficient to evaluate the moral complexity of human enhancement, as they do not deal with the core of the problem of enhancement (Chapter 1). They point out some of the potential consequences human enhancement could have, but do not deal with enhancement per se. Even if forms of human enhancement were safe, distributed fairly, and did not pose a threat to an individual's autonomy, some deeply unsettling problems regarding their use persist. In the cultural imagination, these anxieties are often expressed in science-fiction movies or literature, such as Aldous Huxley's *Brave New World*, a 1932 novel in which the drug Soma is used as a enhancer by humans to experience pleasure, which at least superficially satisfies the

---

[2] Savulescu outlines five basic positions in the debate: "1) Enhancement is morally wrong and should be legally impermissible. This is the strongest negative position. 2) Enhancement is morally wrong but it should be legally permitted. People should be discouraged from employing enhancement technologies, by persuasion, taxation, etc., but they should not be prevented from enhancing. 3) Enhancement is morally neutral and should be legally permitted. This is the position of liberal eugenics. 4) Enhancement is morally right but should not be legally required. Enhancement should be encouraged and facilitated but people should be free to reject enhancements. 5) Enhancement is morally right and should be legally required. This is the strongest position in favor. It requires that people submit to enhancement like they submit to education, fluoridation, or wearing seat belts" (Savulescu, 2009).

moral requirements of safety, justice, and autonomy (Huxley, 2006).[3] But the shallow lifestyle that results from taking Soma is certainly far from morally acceptable.

The anthropological arguments

Acknowledging the limitations of these bioethical normative tools, some have introduced anthropological arguments into the debate that hinge upon concepts of human nature, human authenticity, and/or human dignity. Those arguments have also been used to both argue in favor and against enhancement.

These anthropological arguments, however, fail to recognize that pursuing human enhancement could allow individuals to attain a 'higher' self or their 'ideal' self. Some might strive to become the ideal person they want to be (Chapter 1). Therefore, one variation of these anthropological arguments that has been neglected in the debate appeals to living 'good' human life, or more precisely, appeals to perfectionist elements that are essential to living a 'good' human life, which I refer to in this work as human perfection. This approach differs from arguments of human authenticity, which consider the notion of being true to one's self. The perfectionist approach looks at the vision of the ideal self or the ideal human someone wants to become. Instead of just looking at anthropological arguments telling us what humans are (in which human enhancement would be limited by anthropological assumptions), this approach poses the question of what should we become, given what we currently are. It therefore not only limits enhancement, but seeks to provide some guidance, as it uses an ideal towards which we can strive. This has thus raised the question of whether human perfection could be used as an additional normative tool in the debate.

Working definition of human perfection

It is helpful to distinguish here between the *concept* of human perfection and different particular *conceptions* of human perfection, in the same way

---

3 Additionally, Bonte shows that even though Robert Nozick's famous 'Experience Machine' has "all the preconditions of autonomy, health and fairness met, things can still take a turn for the worse" (Bonte, 2013).

John Rawls differentiates between the concept of justice and different conceptions of justice (Rawls, 1999). The concept of human perfection refers to the view of an ideal human being, including what ideal human properties this human being should and should not have. This concept is action-guiding, as one will try to improve with respect to the vision of perfection embodied in that particular view.[4]

Specific conceptions of human perfection refer to whatever the ideal is. It will perhaps differ from person to person, from culture to culture, and from time to time. It might include having a certain appearance, possessing certain virtues or behaving a certain way, depending on the assumptions that shape the particular vision of human perfection. Nevertheless, as I seek to demonstrate, there might well be some common perfectionist assumptions that could be used in the debate to establish a reference point with which we could better evaluate human enhancement (Chapter 4).

Particular conceptions of human perfection might therefore differ substantially, but ultimately function the same way. They are action-guiding, meaning that they give a goal by providing a vision of an ideal human being. This vision, in turn suggests a trajectory of enhancement for someone to follow to reach that ideal. While some claim to be against perfection in this debate, I will point out that they are actually against a particular conception of perfection, but not necessarily against the concept of perfection in and of itself (Chapter 2). Both proponents and opponents of human perfection often endorse a particular view or some particular perfectionist assumptions to argue either in favor or against enhancement (Chapter 2).

Different conceptions of perfection can be found in the debate, ranging from a full-fledged description of human perfection, to endorsing certain perfectionist elements or assumptions about some particular properties – or set of properties – that would be essential in achieving human perfection. As will be indicated in the following sections, I only subscribe to the latter concept and do not endorse a fixed view of human perfection, or a fully fleshed

---

4   Agar has also recently used a similar concept in which he distinguishes between two ideals of human enhancement that are action-guiding. The first, the objective ideal, refers to the enhancement of specific capcities. The second, the anthropocentric ideal, evaluates human enhancement according to human standards (see Agar, 2014).

out view. Instead, I support a rather flexible and fluid conceptual view that makes room for a plurality of ideas and evolution over time (Chapter 4).

Research question and thesis

This dissertation poses the following research questions. The first question is methodological: in what ways, if any, should the concept of perfection be used in the debate about the ethics of human enhancement? This question leads to the second: if the concept of perfection is useful in the debate, is there a particular conception of perfection that should be used?

The concept of perfection is unavoidable

To answer the first question, I aim to demonstrate that the concept of perfection is unavoidable in this debate. Perfectionist assumptions are also unavoidable, as one evaluates whether an enhancing intervention is morally acceptable according to the reference point ascribed to it by a human being. Furthermore, there are serious shortcomings if one does not take into consideration the concept of human perfection when evaluating human enhancement (Chapter 2). As I will illustrate, such a view is shortsighted and lacks an objective or goal regarding the use of human enhancement. It only gives the appearance of neutrality while implicitly endorsing some perfectionist notions. Therefore, this philosophical debate might miss crucial moral insights if the notion of human perfection is not taken into consideration.

Objections against perfection

Before answering this research question, it is important to respond to the critics within the debate who do not want to address perfection at all. While the rhetoric of perfection appears abundantly in the literature, the notion itself has often been discarded or used negatively, for various reasons.

First, bioconservatives often take the offensive and accuse bioliberals of "seeking perfection." For example, Sandel makes a "case against perfection" (Sandel, 2007), and Kass warns against "biotechnology and the pursuit of perfection" (Kass, 2003). Second, it has been argued that the notion of perfection is unhelpful and unclear as a concept. As Bostrom states in an interview, "I don't think that perfection is a useful concept. There is not necessarily one best form of human existence; perfection might be different for different people" (Bostrom, 2011). Third, critics have argued that the concept of perfection distracts from the real point of the debate. For Buchanan, "The pursuit of biomedical enhancement is not the pursuit of perfection; it is the pursuit of improvement" (Buchanan, 2011, p. 2). This idea is echoed by Vita-More, who states that "Human enhancement and transhumanism aren't striving for perfection, but striving to make us better people" (Vita-More, 2013). Fourth, in a liberal society, it might seem preferable to reject or at least problematize arguments containing perfectionist elements. There is no need to appeal to the perfect or the best, or so it is claimed. According to Brock, it suffices to know what better, but not best, would look like: "Judgments of better, but not best or perfect, are all that are needed to justify selection" (Brock, 2009, p. 272). He further argues that "there is no perfect child, not just in reality, but also even as an ideal. Just as adults' views of the good life are irreducibly different and conflicting, so their views of what a perfect child would be are irreducibly different and conflicting" (Brock, 2009, p. 271).

Responses

These objections are not entirely satisfying. For one, it seems that many participants in this debate do actually resort to perfectionist notions, whether implicitly or not. Paradoxically, even bioconservatives use these notions (see Chapter 2). Both sides of the debate build an ethical argument in a similar manner, and argue that notions of justice, autonomy, and safety are not enough to build a case against enhancement. They further build a case against enhancement by presenting a vision of the ideal human, or at the very least by outlining some of the perfectionist elements this ideal human should have. For them, human enhancement is a threat to these perfectionist notions, and so it must be banned. Buchanan also recognizes this paradox of both accusing opponents of pursuing perfection while at the same time relying on some perfectionist notions. He suggests, though, that the notion of perfection be done away with (Buchanan, 2011)[5]. Nevertheless, it seems

---

5   Buchanan writes, "Ironically, while sternly criticizing enhancement as the quest for perfection, the President's Council pins one of its chief arguments against enhancement on an unspoken assumption that anything that departs from what they assume to be the best sort of human procreative activity is wholly unacceptable, demeaning, and subhuman" (Buchanan, 2011, p. 127–28), and continues,

that bioliberals do implicitly built their arguments on perfectionist notions as well. Some bioliberals position the value of autonomy as the ultimate ideal to strive towards (Harris, 2007). Under this view, autonomy takes on some perfectionist assumptions as an essential precondition to leading a good human life. Thus, instead of eliminating the notion, it has highlighted the need for additional clarification. As Keenan rightly explains: "The problem lies not with the question of whether we should pursue perfection, but rather what perfection we are pursuing" (Keenan, 1999, p. 104).

Second, if the notion is unclear, it does not imply that the point has been missed, but again emphasizes the need for clarification. The notions of autonomy, justice, and dignity are also contested, but instead of getting rid of them altogether, many attempt to provide further clarification.

Third, it is not obvious to all that the notion of perfection misses the point of the debate. As Walker writes:

> One reason to use the term 'perfection' rather than merely 'improvement' is that it clarifies the axiological structure: 'perfection' is an end in itself. To say that our goal is merely to 'improve humans' is to invite the questions of 'how far?' and 'to what end?' (Walker, 2002)

It is also unclear how one can judge whether something is better or not without resorting to an ideal reference point. As I will elaborate later in Chapter 3, two ways of understanding what 'better' means have often not been recognized in the debate. The concept can be defined in the following ways: i) better in comparison to a former state (backward-looking), or ii) better according to an ideal or perfection (forward-looking). In the latter definition, the concept of perfection becomes helpful, as it provides another reference point by which one can evaluate whether something is better or not.

Fourth, in order to respect the plurality and diversity of ideas and people in a liberal society, it is supposed to be preferable to consider only what is better, not what is best or perfect. However, only looking at what is considered better does not resolve everything and the question of what better means, according to whom, still remains. Some bioliberals do not want

---

"Here we encounter again the irony noted earlier: those trying to make a case against perfection do so from the standpoint of a very dubious perfectionism" (Buchanan, 2011, p. 134).

to take into consideration perfectionist arguments, as they are based on particular conceptions of perfection. These conceptions might hint at or reveal certain preferences that may work against or devalue the choices of individuals living in a pluralistic and liberal society (Chapter 2). I argue, however, that we do need to take the concept of perfection into consideration. Not only is the concept of perfection unavoidable in the debate (Chapter 2), it is also from these perfectionist assumptions that human enhancement is assessed as being morally problematic or sound. Therefore, the concept of perfection should not be so quickly disregarded, and it calls for our attention and further inquiry.

Which conception of perfection should we use?

After defending the position that the concept of perfection is unavoidable, I will ask which conceptions of perfection should therefore be used (Chapter 4)? I will argue that the best conception of human perfection ought to fulfill a certain number of criteria. It should be an objective ideal, but not one that holds a fixed view of what a human being ought to be. This flexible approach is meant to allow room for pluralism and political deliberation, while also providing guidance – and not only restriction – relating to the practice of enhancement (Chapter 4). Martha Nussbaum's capabilities approach fulfills these requirements and can therefore be used to morally assess human enhancement, in addition to other bioethical standards such as safety, justice, and autonomy (Chapter 5).

In order to arrive to this conclusion, I first challenge and invalidate problematic conceptions of perfection, after which then I examine the content, justifications, and use of different conceptions of perfection in the debate. Regarding its content, I use Walker's categories to distinguish between type-perfection and property-perfection. Type-perfection is "the thesis that those individuals who best realize the essential properties of the individual's type or species best exemplify the ideal of perfection" (Walker, 2002). Property-perfection is defined as the "thesis that those individuals who best realize some property or properties best exemplify the ideal of perfection" (Walker, 2002). I argue that property-perfection is not plausible, as it falls back onto type-perfection. Therefore, addressing type-perfection cannot be avoided.

Two possibilities are available for justifying type-perfection: the objective or a subjective approach. I argue that the better option is the objective approach, as the subjective approach does not stand up to further scrutiny, and entails the additional problem of introducing objective perfectionist assumptions.

Regarding its usage, type-perfection has previously been used to ban or limit human enhancement. I argue here, however, that this does not have to be the case. A conception of perfection that is only used to restrain or limit human enhancement is not sound, as it is unclear why this ideal is only used in the negative, when it could also help provide positive guidance for human enhancement.

I then take a positive approach and suggest which particular conception should be used in the debate. By applying Nussbaum's capabilities approach (Nussbaum, 2011), I show how the capabilities approach fulfills the requirements I have been looking to fulfill. I put forth the argument that human enhancement is morally acceptable if it does not undermine a human's core capabilities, but rather maximizes them in a holistic fashion. For Nussbaum, the central capabilities of the human being include: life; bodily health; bodily integrity; senses, imagination, and thought; emotions; practical reason; affiliation; other species; play; and control over one's environment (Nussbaum, 2011).

Using Nussbaum's capabilities approach as an objective framework demonstrates the necessity of having a public debate to discuss how the central capabilities are being modified with human enhancement. This framework also allow for a certain degree of both diversity and unity within human enhancement. Each human capability is open for public discussion; which would in turn encourage the formation of a non-fixed ideal, yet an ideal that is both objective and flexible. This ideal should be able to change overtime and depend on particular circumstances.

This approach offers several advantages. First, it helps us to look at how a particular form of enhancement can extend, add, remove or decrease a specific capability, thereby affecting the central capabilities as a whole. It forces us to look at how life-extension, for example, would actually affect the central capabilities as a whole – not just one capability. So while a particular capability can be improved while enhancing, it might be detrimental to another central capability. For instance, a highly intelligent enhanced person might become unable to relate to his peers, and this social isolation could have a negative impact on his physical health. The central capabilities form a whole, and each individual part within that whole may come into conflict or tension with one another. This approach forecloses an overall assessment that helps us think of a human enhancement in terms of its impact on the central capabilities as a whole. The existence of a certain harmony within the central capabilities is a kind of prerequisite needed that needs to be in place when considering whether a particular human enhancement is worth pursuing.

However, adding or removing a specific capability is not necessarily wrong, but it would become morally wrong if this modification would have a negative impact on the central capabilities as a whole. This premise would allow some forms of human enhancement, while clearly banning others. The important point here is that the central capabilities as a whole should not decrease, but should be maximized in a harmonious way.

Second, this approach provides both guidance and limitation to human enhancement by establishing a reference point that can be used to illuminate a certain direction to aim for the enhancement to follow.

Third, the ideal human that individuals are striving towards when using human enhancement should at a minimum retain the ten central capabilities; a form of human enhancement should harmoniously maximize those capabilities as a whole. In a practical context, other specific capabilities are added to the central capabilities so that human enhancement can be evaluated within this framework. For example, if one wants to become a great basketball player, then she should have some other specific capabilities. Those specific capabilities, however, should not limit the central ones. This provides a safeguard against what might appear to be a human enhancement, but may actually instead be a diminishment. If a particular capability is increased (e.g. becoming taller to play basketball), this might affect negatively central capabilities, (i.e. becoming so physically altered or unusual appearing that it makes it difficult to relate [capability of affiliation] with the world).

Outline

This research is divided into two parts. The first part is a methodological and argues that the concept of human perfection is inevitable in the debate.

The second part is substantive and shows what requirements a conception of perfection should fulfill before defending a particular conception of perfection based on Nussbaum's capabilities approach. After a *clarifying* overview and an examination of different definitions and normative tools used in the debate, I provide a *critical analysis* of the use of the concept of perfection by both bioconservatives and bioliberals. This critical analysis is followed by some *conceptual work*, which demonstrates that the notion of human enhancement and human perfection are both intertwined. Finally, I conclude this study with some *constructive work*, and lay out the requirements a conception of perfection needs to fulfill in the context of this debate. I then argue in favor of using one particular conception of perfection based on the capabilities approach.

To answer the research question, the project is divided into four papers, which are presented here as five chapters. The first chapter outlines different definitions of human enhancement and defends a qualitative definition. I then demonstrate how the current bioethical notions of safety, justice, and autonomy are limited when it comes to evaluating the morality of human enhancement, and thus suggest that the notion of human perfection be considered as an additional normative tool. The second chapter shows how bioconservatives paradoxically endorse perfectionist notions, while accusing bioliberals of seeking perfection. I then raise the question of whether bioliberals can really avoid speaking of human perfection. In the third chapter, I outline how the notion of improvement can be understood in two ways: a backward-looking approach and a forward-looking approach. I illustrate that a backward-looking approach is limited and that a forward-looking approach is better suited to evaluating the morality of human enhancement. This discussion demonstrates how the notion of human enhancement is conceptually related to the notion of perfection. In other words, in order to make a qualitative comparison between two states of affairs, having a third additional ideal reference point is needed. In the fourth chapter, I challenge and ultimately reject problematic conceptions of human perfection. Finally, in the last chapter, I defend a particular conception of human perfection based on the human capabilities approach of Nussbaum in which I argue that the capabilities approach can be used as a framework to evaluate human enhancement.

Implications

This work has several implications. First, this thesis takes a slightly different approach than existing anthropological arguments found in the debate, as it does not look at what our human nature is, or ought to be, but rather considers what type of human we are hoping to become with the help of human enhancement. It also differs from arguments regarding human authenticity because, as previously suggested, the aims of human enhancement are more than just becoming who I truly am, and are more so about becoming who I could ideally be.

Second, this examination not only helps clarify the concept of perfection; it also helps answer the question "better according to whom?" by acknowledging that the notion of improvement itself can be understood in two distinct ways: improvement over a former state, or improvement towards an ideal.

Third, this approach provides an additional normative tool to help evaluate the morality of human enhancement and adds to the notions of justice, safety, and autonomy. It encourages us to take a more holistic approach, challenging us to think about how altering one capability may influence other capabilities. In this sense, one does not just enhance one's life expectancy; this enhancement would also have an impact, whether positive or negative, on the individual capabilities of affiliation, health, and bodily integrity, among other central capabilities. It also views the human in relation to other humans, and not just as an individual. This approach thus encourages us to evaluate the morality of human enhancement in a more holistic fashion, and not only look at the enhancement of specific abilities without taking the whole human into consideration. This more holistic framework in turn encourages a more comprehensive evaluation of the key issues at stake in the debate.

Fourth, this view offers a far more objective and communal approach to the ethics of human enhancement than a typical subjective libertarian ethos. It differs from both the bioconservative and bioliberal approaches in a number of key ways: Bioconservatives use perfectionist assumptions only to limit human enhancement, while this study uses the same assumptions to both guide and limit human enhancement, and bioliberals deny the usefulness of perfectionist assumptions, while this works defends that their use is inevitable. In addition, this research is distinct from a transhumanist approach because it focuses solely on *human* enhancement. Type-perfection of the ideal human sets limits to the practice of enhancement. As I demonstrate, however, those limitations are not fixed, but are fluid and flexible. Furthermore, a transhumanist or even posthumanist approach would also include some sort of limitation to enhancement, since the concept of type-perfection cannot be avoided. In this case, it would be transhuman-type-perfection or posthuman-type-perfection. Nonetheless, it would be some sort of type-perfection.

Fifth, the capabilities approach provides a helpful framework that can be used to assess each particular enhancement on its own terms by looking at how it would affect the central capabilities. Here, it is also important to consider the complexity of enhancements in theoretical individual cases. For instance, there may be some circumstances in which decreasing intelligence could be seen as an enhancement, while in other situations it may be morally unacceptable.

Sixth, this approach or methodology could also be used for other types of enhancement, such as posthuman enhancement or animal enhancement. One would have to define the central capabilities that are essential to both of these types and then see how enhancement could be implemented. Even in these nonhuman realms, not all forms of enhancement will be permissible, since perfectionist notions will be both limiting and guiding. For example, one might include the capability of flying among the central posthuman capabilities, which would then make it possible to use the central posthuman capabilities as a frame of reference to evaluate the morality of posthuman enhancement.

Scope and limitations

The project seeks only to answer the research question stated above: in what way, if any, should the notion of perfection be used in the debate regarding human enhancement? It does not seek to produce an exhaustive ethical framework that could be used to answer every question or issue pertaining to human enhancement. The research question is answered by demonstrating that the concept of perfection is unavoidable and that a particular conception of perfection can be found in the capabilities approach.

The project adds to the current debate and to other arguments that examine autonomy, justice, safety, authenticity, and human nature. The research presented here is limited in scope by considering only HE and not at other forms of enhancement such as posthuman or animal enhancement.

Ultimately, this research has remained conceptual, but in future research, it will be necessary to look at different forms of enhancement separately (e.g. cognitive enhancement, sport enhancement, etc.) and to apply this conceptual work to particular cases. This current project responds to the debate from a general perspective, similar to the ways other prominent authors such as Kass, Sandel, Fukuyama, and Agar, have approached the discussion.

Chapter 1: Living up to the ideal human

Summary

Two major problems persist in the debate regarding the ethics of human enhancement. First, there is a general lack of discussion and consensus on a definition of human enhancement. Second, the commonly used bioethical principles of justice, safety, and autonomy are together insufficient to assess the morality of human enhancement. This chapter attempts to define these problems and propose a possible solution. Defending a qualitative definition of human enhancement, the authors suggest examining 'perfectionist notions' of what it means to live a good human life in order to provide additional normative tools that can be used to evaluate the morality of human enhancements. This chapter clarifies the existing debate and helps to move it forward by arguing that defining the characteristics of the ideal human life as the goal of human enhancement can help judge the morality of a given form of human enhancement.

Introduction

One of the most debated topics in bioethics in recent years has been the ethical issues related to human enhancement. While many articles have been written on this topic, two major problems still persist. First, there is still no real consensus on the definition or meaning of 'human enhancement,' which complicates discussions in the debate. Second, the commonly used bioethical principles of *justice*, *autonomy* and *safety* are together insufficient to assess ethical issues related to human enhancement. This chapter aims to provide answers to both problems by outlining and analyzing various attempts to define 'human enhancement.' I then suggest that the bioethical standards of *safety*, *justice* and *autonomy* are together insufficient to normatively frame the ethical issues pertaining to human enhancement. Finally, I suggest a novel way of assessing ethical aspects of human enhancement by using a qualitative definition of human enhancement and perfectionist notions of living a good human life.

The conundrum of defining human enhancement

Human enhancement is tacitly defined as technologically modifying human bodies and/or minds by combining medical science with emerging technologies (such as nano- or biotechnologies) to affect individuals' cognitive (e.g. memory, intelligence), emotional (e.g. happiness, self-confidence) or physical (e.g. sleep requirements, endurance) functioning. This kind of consensual agreement on what could be considered human enhancement, though, does not constitute a definition of the concept. Interestingly, when examining the literature addressing ethical issues related to human enhancement, very few authors have explicitly defined the concept (see e.g. Buchanan, 2011; Menuz et al., 2013; Savulescu, 2006). Generally speaking, authors only offer an implicit definition of the concept that appears in their argumentations. However, it is possible to separate these implicit or explicit ways of defining human enhancement into different categories (see e.g. Chadwick, 2008; Menuz et al., 2011).

In this section, I will therefore discuss three approaches that have been used to understand human enhancement: i) the beyond therapy approach; ii) the quantitative approach; and iii) the qualitative approach. I will argue that while the qualitative approach is the most plausible definition of human enhancement, they all have some substantial shortcomings, which I will resolve in the final part of this chapter by using a recent definition suggested by Menuz et al.

Beyond: the 'beyond therapy' approach

Some commentators have defined human enhancements as medical interventions that do not attempt to cure. Such a view can be defined as 'beyond therapy' or 'outside the scope of therapy' (President's Council, 2003). According to this view, the goal of such technological interventions is not to heal, but to go 'beyond health' or outside the realm of therapeutic intervention (e.g. cosmetic surgery). One influential interpretation of this view asserts that healing consists of maintaining – or returning back to – a normal range of human functioning. Enhancements, in contrast, are interventions that surpass a normal range of functioning.

This definition of human enhancement has been controversial due to the difficulties related to its application, mainly because it is almost impossible to establish or agree upon what 'normal functioning' in fact is. Similarly, doubts have been raised about the ability to objectively define health and disease (Buchanan et al. 2001). Recent discussions have pointed out that the distinction between therapy and enhancement cannot be clear-cut (Daniels, 2000; Harris, 2007; Menuz et al., 2011). Moreover, this definition of human enhancement suggests in itself an entirely negative approach of the concept: the 'beyond therapy' definition implicitly posits that there is a 'good' (i.e. therapy) *versus* a 'transgressive' (i.e. human enhancement) way of practicing medicine. In other words, human enhancements are ethically suspicious results of technological interventions, while the same interventions used for the treatment of disease are ethically acceptable. This way of defining human enhancement is often built upon value judgments that lend a negatively charge assumption to the concept itself. In addition, defining human enhancement as the result of biotechnological interventions that extend beyond the bounds of medical therapy might lead to counter-intuitive judgments. For instance, the use of non-therapeutic drugs to reduce the need for sleep might be considered a form of human enhancement, and by that same logic, the use of non-therapeutic drugs to commit suicide would also be classified as such under this definition.

There is an essential, but contested distinction between therapy and human enhancement, making it difficult to define what is – or is not – human enhancement.

More: the 'quantitative' approach

Some authors have defined human enhancement as an add-on to existing human characteristics (Chadwick, 2008, 2011). Defining human enhancement in this way is characterized as a quantitative approach in which any technological intervention that increases specific characteristics – such as cognitive capacities or physical abilities – can be regarded as human enhancement. While the 'beyond therapy' approach has been the most frequently used in existing literature on the subject of the ethical issues in human enhancement, the add-on definition is seldom used. There are two major reasons to explain this. First, enhancement is viewed by many as an improvement in quality, not necessarily in quantity. Many commentators consider the qualitative dimension of human enhancement as crucial (e.g. Buchanan, 2011a; Buchanan, 2011b; Harris, 2007; Heilinger, 2010). Second, it is also incorrect to assume that a given human enhancement necessarily requires an addition. For instance, there are many cases from surgery in which removal rather than addition underlies enhancement, as illustrated by a nose or breast reduction for cosmetic purposes (see e.g., Earp et al., 2014).

Better: the 'qualitative' approach

Some have argued that human enhancements are technological interventions that have the potential to improve the life of person being modified. This qualitative way of defining human enhancement has recently become one of the predominant views in the debate (e.g Buchanan, 2011a; Buchanan, 2011b; Heilinger, 2010). On the one hand, this definition contrasts with the 'beyond therapy' approach because it assigns a default positive characterization to human enhancement. On the other hand, it diverges from the 'quantitative approach' because it refers to a qualitative idea of improvement rather than relying purely on the quantitative idea of an addition. While I agree that such a definition is a better definition of human enhancement, this still leaves the unresolved questions of who decides whether a given technological intervention has the potential to improve life, whether the result of a given intervention can – or cannot – be considered human enhancement. These questions will be partially answered in the last section of this chapter.

Limitations of ethical tools used to evaluate human enhancement

In addition to the difficulty of defining human enhancement, the many attempts to ethically evaluate human enhancements remain a conundrum. To this point, such ethical evaluations have mainly been based on bioethical principles, and notions of safety (Annas et al., 2002; Fukuyama, 2003; McKibben, 2004; Mehlman, 2009), justice (Buchanan et al., 2001; Caplan, 2009; Habermas, 2003), and autonomy (Agar, 2004; Buchanan et al., 2001) have been central to the debate. However, while a given human enhancement may meet the requirements of one of these principles, this does not necessarily mean that it is ethically acceptable. For example, applying this definition to Aldous Huxley's *Brave New World* demonstrates that even though the happy-pill 'Soma' meets the requirements of safety, justice and autonomy, something quite unsettling about the entire practice remains (Bonte, 2013).

Actually, the established bioethical principles appear not able to cover all the ethical complexities that human enhancement may raise, such as, for example, threats to human nature (dehumanizing), the desire for human beings to play God (hubris), or the alteration of dignity and authenticity (alienation).

In this section, I will outline and criticize the three commonly used bioethical principles and argue that while they may all be necessary, they are together insufficient to evaluate the morality of human enhancement. I will then move on to consider other philosophical arguments and then argue for my own in the last part of this chapter. I believe that the inadequacy of the bioethical approach is due at least in part to the absence of a precise definition of what constitutes a human enhancement and for this reason, the last part of my analysis provides a specific definition of human enhancement.

Safety

The principle of safety suggests that we should refrain from forms of human enhancement that are not considered safe. While this position may be appealing at first, the argument of safety is not as straightforward as would initially appear.

There are indeed situations in which an individual's quality of life might be so poor that the risk of undergoing a certain technological modification that has the potential to improve her condition makes that risk acceptable and even commendable. In other words, subjectivity and objectivity come into conflict when evaluating the safety of a given human enhancement: the criteria by which to assess a human enhancement as safe or unsafe are based largely on the individual's perception of the inherent risks and benefits. Importantly, when an individual assesses whether something will be beneficial for her, she needs to be aware that there may be unpredictable side effects, such as developing an addiction to technological modifications, or having to undergo further procedures to fix a bug or update the technological modification.

With respect to human enhancement, one should bear in mind that unintended consequences may be inevitable (Fukuyama, 2003; McKibben, 2004). For instance, conflicting situations may arise between those who have been enhanced and those who have chosen not to undergo modification, whether for personal or financial reasons. Others worry that enhancing humans may lead to a new kind of eugenics, in which parents might be pressured or even forced to enter into a competition to produce the best possible child (Sparrow, 2011). By attempting to improve humankind through technological modifications, we may end up discriminating against certain traits or populations who do not display a desired trait. There is also a risk of diminishing human diversity: as certain traits become desirable or undesirable, passive and/or active coercion may compel populations to adhere to social norms. In this regard, some philosophers theorize that the tensions between technologically enhanced and unenhanced individuals may lead to an increase in individual responsibility, coupled with an erosion of solidarity between enhanced and unenhanced individuals (Sandel, 2007).

The principle of safety raises interesting conflicts regarding the ethical issues related to human enhancement. Used alone, this principle has too many weaknesses to effectively evaluate the ethical issues relevant to human enhancement. However, it offers a glimpse of the many difficulties that do arise when trying to develop and use a framework for ethically evaluating human enhancements.

Justice

Technological interventions aimed at enhancing human beings could potentially reduce the gap between those who have, and those who have not. Some have therefore suggested that it could be used to help make societies more equitable (Buchanan et al., 2001). Indeed, one may wonder whether there are reasons to accept natural injustices and reject the use of technological interventions in order to compensate for these inherent discrepancies. Such technological interventions, however, might actually widen the gap between those who have and those who have not (McKibben, 2004), bringing unfair advantages to technologically enhanced individuals (Mitchell, 2009). Some have even suggested that such enhancing interventions might be perceived as cheating (Schermer, 2008): is taking a cognition-enhancing pill before an important exam, for instance, fundamentally unfair?

It seems that ethical issues related to human enhancement are not rooted in the technologies themselves, but in their access and distribution. Philosophers such as John Harris wonder whether a given human enhancement should be judged as unethical on the basis that such enhancement will not be affordable or attainable for the vast majority of citizens in most societies

(Harris, 2007). For Harris, the argument is weak: generally speaking, new technologies are first used only by a minority of individuals. If the technology then proves to be beneficial, the prices usually go down, as the mass-produced technology becomes more accessible (Harris, 2007). Some have called Harris' point naïve because even though technologies do tend to become cheaper when mass produced, this does not necessarily have a substantial impact on the accessibility of these technologies to members of a given society. Following this line of argumentation, technological interventions leading to human enhancement would not help to overcome or solve inequalities (Zylinska, 2010), and they might in fact worsen the existing social gaps between individuals in many societies.

In this context, one may wonder whether human enhancements themselves actually raise ethical issues, or whether such issues are embedded in the economic structure of our societies and not necessarily in human enhancement itself. If this is true, the issues raised here would pertain to distributive justice, not human enhancement per se.

Autonomy

According to some, autonomy is a facet of life that could be increased or diminished by human enhancement. On the one hand, some forms of human enhancement may lead to an increase in an individual's *functional* autonomy by giving them a greater degree of independence to move and act more freely (Buchanan et al., 2001). For example, a given human enhancement might help some people to be more independent in their work or daily routine. Some human enhancements could therefore be ethically justifiable because they serve to increase people's autonomy. On the other hand, an individual's *decisional* autonomy may be threatened by the social pressure to conform, which would result from the widespread use of technological modifications (President's Council, 2003; Menuz, forthcoming). This begs the question of whether human enhancements would be chosen freely or imposed by other societal standards. This potential threat to autonomy becomes even more important when issues of enhancing children or embryos come into play. Should parents, for example, be allowed to try to technologically modify their children because of their procreative autonomy (Harris, 2007)?

Human enhancement could also have a negative impact because it threatens the preconditions for autonomy, which can be found in the authentic self or personal identity (DeGrazia, 2005). However, for some, autonomy itself, understood as reasoning ability, could be increased through human enhancement (Schaefer et al., 2013). At first the principle of autonomy alone does not seem to be enough to solve the ethical issues raised by human enhancement. While whole principles are limited to providing normative frameworks for understanding human enhancements, they do help shed light on the complexity regarding the ethical evaluation of human enhancements.

Anthropological arguments

Ethical arguments based on the anthropological and philosophical notion of *human nature* have recently been introduced into the debate on human enhancement (Heilinger, 2010). Some hold that tampering with our shared human nature might undermine the equal status of individuals or threaten the inherent worth of human beings, which is understood to be the basis of human rights (Fukuyama, 2003). In the same vein, others have argued that our lives might become dehumanized and meaningless if we alter inherently human traits like finitude and vulnerability (President's Council, 2003; Kass, 2003; Parens, 1995; Sandel, 2007). In contrast, some have argued that we possess human dignity, which encourages us to use technological intervention to enhance ourselves (Bonte, 2013).

Other critics have labeled such arguments dubious and thus unhelpful because they often rest upon controversial metaphysical or theological assumptions. Accordingly, these arguments thus commit a naturalist fallacy (Buchanan, 2011b) or fail to be appeal to a diverse audience. To avoid this natural fallacy, Bostrom defends posthuman dignity (Bostrom, 2005). However, even within a liberal framework, appeals to human nature can be shown to be an important aspect in debates about human enhancement (Heilinger, 2010; Schramme, 2002).

While these lines of argumentation are interesting and do contribute to the debate on human enhancement, they still miss the point that human enhancements might not be used to discover a true self (human authenticity) (see e.g., Erler, 2012; Levy, 2011), but to create an ideal self informed by different ideals individuals have regarding what constitutes a good (human) life. Some arguments of human authenticity and human dignity (Kass, 2004) also miss this point. When used, those notions fail to acknowledge that ideals can also guide actions, not only restrict them. This is why some have introduced arguments based on perfectionist notions of what it means to live a good life (Roduit et al., 2013), hoping to enlighten the debate with another type of approach. I will therefore focus the rest of the discussion on these perfectionist assumptions by first putting forth a specific definition, and then analyzing whether these assumptions can be helpful in this specific debate.

Living up to the ideal human: evaluating human enhancement

Having now outlined two major problems in the debate on the ethical issues related to human enhancement – first, the lack of a consensus regarding a definition, and second, the limitations of its current ethical evaluation – I will now suggest a new way of approaching both issues. In this section, I will lay out a perfectionist approach of what it means to live a good human life that can be used to frame ethical issues related to human enhancement.

Defining human enhancement as a qualitative improvement can provide a solid foundation for building a framework to ethically assess human enhancements. Recently, a new angle has been proposed and according to the authors (Menuz et al., 2011), defining human enhancement is not possible without referencing personal and subjective perceptions, which are themselves influenced by socio-cultural factors (e.g. political and social norms, rules, values, environmental factors, passive coercion, unconscious goals, and/or statistically defined attributes, considered within a given society in a given historical period of time). Such perceptions constitute the 'personal optimum state,' that is, an individual's ideal of life. Under this definition "each individual has to determine for herself/himself, based on her/his personal optimum state, whether the outcome of a given technological intervention can be described as human enhancement or not" (Menuz et al., 2011).

Here, I continue therefore exploring the ethical justifications for using human enhancement to reach an optimum human state. Indeed, in order to evaluate human enhancement, a reference point is needed because a qualitative improvement can only be interpreted as such in reference to something. As previously stated, notions of safety, justice, and autonomy are not the only reference points available to us; considering what would the ideal human consist of is another one. Put more precisely, since it is not possible in a pluralistic society to agree on such ideal, one can theorize what individual characteristics an ideal human would have. Taken together, these traits would then provide a reference point to be used as a normative tool in order to morally assess whether an enhancing intervention is acceptable or not. The difference here is that this reference point also need not be normal human functioning or human authenticity, but human ideals, since human enhancement can be used not only to become a true or an authentic human, but an ideal one.

Two views of improvement that have not often been used in the debate deserve our attention. The first is the *backward-looking view*, in which human enhancement can be evaluated according to a former reference point. For Harris, a human enhancement is "by definition an improvement on what went on before" (Harris, 2007, p. 9). Second is the *forward-looking view*, in which human enhancement is evaluated according to an ideal reference state, or what has been called perfectionist notions of what it means to live a good life (see e.g. Roduit et al., 2013; Walker, 2002). Consider the case of someone who wants to use a 'height enhancer': according to the backward-looking view, the enhancement is evaluated on the basis of someone growing taller, say from 160cm to 165cm. In retrospect, the person gained five centimeters in height and the intervention would thus be perceived as an enhancement. On the other hand, the forward-looking view would look at ideals that can be used as a reference point to evaluate whether an intervention had been successful or not. Here, if someone desires to become a professional basketball player, the height increase will be measured as an enhancement according to the ideal height for playing basketball. As outlined elsewhere (Roduit et al., 2013), the backward-looking view has some serious shortcomings because it fails to take into consideration the forward-looking view as well. It neglects the reality that that some forms of human enhancements may have a direction or goal in mind, as individuals enhance 'towards' an ideal, such as the ideal basketball player. In short, the backward-looking view lacks an ideal, and is thus (literally) shortsighted. The forward-looking view is therefore a better perspective, as it resolves these shortcomings (Roduit et al., 2015).

This second possibility of assessing human enhancement introduces perfectionist elements of what it means to live a good human life. These perfectionist elements can be used to evaluate whether an intervention counts as an enhancement or not. In comparison to a strictly backward-looking approach, this view offers additional insights into the debate that should not be ignored. Different ideals of what it means to lead a good human life can be used to interpret whether an intervention can be thought of as an enhancement or not. In addition to this, it is helpful to look at this debate directly from this larger perspective because the concepts of human dignity or authenticity do vary according to various views of the good human life.

These perfectionist perspectives, or this forward-looking view, can be used in two different ways. In the debate, Walker identifies two alternatives and distinguishes between *type* and *property* perfection. Type-perfection is "the thesis that those individuals who best realize the essential properties of the individual's type or species best exemplify the ideal of perfection" (Walker, 2002). Property-perfection is the "thesis that those individuals who best realize some property or properties best exemplify the ideal of perfection" (Walker, 2002).

Some problems do arise with the property-perfectionist view, though, as it ultimately falls back into type-perfection, because a property is always related to a type. For instance, while some may desire to increase their intelligence, recognizing intelligence as an objective good, the property of intelligence will always be embedded into a type, regardless of whether this type is human, sub- or post-human. While the property-perfectionist view has the advantage of looking at different properties one can enhance, it neglects the bigger picture and the fact that these properties belong to a given or a chosen type. The type-perfectionist view is therefore the most appropriate.

Type-perfection can also be further distinguished between objective type-perfection and subjective type-perfection. On the one hand, a type is given or agreed upon objectively (e.g. the type human, monkey or chair). On the other hand, a type could also be defined subjectively. Here, we could image a human enhancing towards something so unique – according to her own subjective desires – that it would not belong to any other type anymore. However, the latter option is not plausible when one speaks of human enhancement, because it is the human – as a particular objective type – that is ultimately being enhanced.

We are therefore left with the objective type-perfection option in which a certain type can serve as reference for what it means to lead a good human life. Indeed, this type is made of essential properties that would exemplify what the ideal type is, which in our case is the ideal human.

While for some this type of approach may seem paternalistic and problematic in a pluralistic society, these problems can be avoided. First, while it is true that members of diverse societies have difficulties on agreeing about virtues or objective goods, we can still have public discussions and come to an agreement regarding what essential properties are necessary in order to lead a good human life and to be considered 'human.' Even in our liberal societies, we do generally agree that some types of living are better than other.

Second, some would claim that autonomy alone would be a better way to assess the morality of human enhancement, as it would avoid paternalistic tendencies. However, we already limit people's autonomy in relation to other goods, such as education. Autonomy is sometimes limited for a time so that one will eventually become more autonomous or will have the essential properties to live an autonomous life. We also limit people's autonomy if it poses a threat to others. We can therefore seek to reach consensus in public debate about the essential human features or capabilities necessary for human to live a good and vibrant life.

In order to speak of this autonomous human being, we also need to have a human, but this chapter does not outline what a particular ideal human or ideal humans would be like. A plausible answer to this question can be found in the work of Sen or Nussbaum, for example (Sen, 2009; Nussbaum, 2011). As mentioned elsewhere, different ideals may have different content, but these ideals are used in the same ways: they are action-guiding and help one to make a decision concerning the direction the individual wants to move towards by enhancing a capability (Roduit et al., 2013). These ideals can either be given or agreed upon. For some, the type 'human' is a given, whether ordained by nature or god (the essentialist approach), while for others it might be socially constructed (the existentialist approach). For our purposes, whether it is one approach or the other does not matter here, as the ideal has the same function in both cases: it is action-guiding and is thus able to guide the enhancement project.

To move deeper in the debate, additional discussions about what type of humans we want to become is necessary for us to know and understand which type of human we seek to become. In other words, knowing what the ideal human(s) is should enable us to use enhancement in a morally acceptable way, as we would then use this particular type to orient and guide the forms enhancement takes. Asking ourselves what the essential features needed for living a good human life are will help build a framework that can help guide human enhancement. As described by Baertschi:

> [N]either our desire for enhancement nor our concerns about personal identity can be properly understood without referring to an explicit or implicit ideal: *the ideal of the person we want to be*. This ideal is an essential part of our conception of the good life, because a good life is a life we want to live, as the person we want to be. (*emphasis mine*, Baertschi, 2009, p. 39)

Advantages of this approach

This approach holds a number of different advantages:

1. It differs from the notion of authenticity (who I truly am) by putting more emphasis on the ideal person one strives to be (who I want to become). Because human enhancement is about more than just uncovering who we truly are, it is also a means for realizing and becoming one's ideal self.
2. The perfectionist approach adds a consideration of what it means to lead a good human life to a strict notion of autonomy, but refuses to use autonomy as the only possible ideal, as it has been the case with some bioliberals (Roduit et al., 2013; see e.g. Harris, 2007). Indeed, for some the ideal being they strive towards is an autonomous being/person, but this is not the case for everyone. With the perfectionist approach, a wider range of ideals is acceptable. Recognizing, however, that not all views of the good human life are morally acceptable, a public discussion becomes necessary to help define what type of human we would or would not accept in our respective societies.
3. This approach also points out that some (societal) ideals influence the choices of individuals who wish to enhance in a certain way. One could say that some sort of enhanced account of autonomy is defended here, taking into account that participants in the debate have perfectionist assumptions regarding the good human life, which influences whether they

view an intervention as an enhancement or not. This view encourages individuals to be self-reflecting not only regarding the ideals they are striving towards, but also some underlying societal ideals that probably influence them to consider using a given enhancement in the first place. It points to more psychological questions: why does an individual wish to enhance (or not) in a particular way?
4. This approach also acknowledges that human enhancement is not only an improvement over a former state, but also an improvement leading somewhere – towards the ideal self. Human enhancements are therefore a means to an end. Problems are therefore not to be found in the means, but in the ends.
5. Finally, while not identifying anything intrinsically morally wrong with human enhancement, this view still recognizes that the ideal influencing some forms of human enhancement might be morally unacceptable if this ideal involves hurting others. Similarly, some kinds of human enhancement might be unwise or plain irresponsible. Here, individuals judge whether an intervention is an enhancement for their lives by evaluating the advantages of the interventions for themselves. This view seeks to avoid any type of social coercion or pressure.

This approach can therefore add to the current debate and theoretical tools used in the debate to assess human enhancement. Needless to say that, bioethical standards such as justice, autonomy, safety and human dignity, while limited, still do have a role in this debate. My approach adds another tool for evaluating human enhancement so that an ethical evaluation can be made in a more exhaustive way.

Conclusion

One of the major obstacles to framing the ethical issues surrounding human enhancement is the lack of a precise and shared definition of what constitutes human enhancement. Moreover, the many definitions that have been proposed – whether implicitly or explicitly – do not take into account the complex contexts in which human enhancements may occur. Both elements partially explain why bioethical principles, such as autonomy, justice and safety, have failed, so far, to provide satisfying answers in this debate, especially regarding issues such alienation, hubris or dehumanization.

By basing my work on a qualitative definition of human enhancement, I have developed a preliminary conceptual framework of the ideal human being to help investigate the ethical issues related to human enhancement. Human enhancement is not only a subjective concept where individuals choose what an enhancement is for themselves; it is also objective because it concerns the 'human.' Adding to other bioethical notions, I have argued that the ethical evaluation of a given human enhancement also needs to account for humans, who make ethical judgments of a given human enhancement in accordance with the ideal toward which they choose to enhance (perfectionist assumptions). For this chapter, I have purposely neglected to outline what those human properties essential for an ideal human might be, or what it might mean to live a good human life. These big questions require not only a great deal of space and consideration, but also a public discussion, both of which are outside the scope of this chapter. The last chapter of this book will outline a possible solution.

The suggested definition and type-perfectionist view given here enable us to acknowledge the importance of the concept of the ideal human. Ultimately, the question is not whether we will enhance, but towards which view of the ideal human we will enhance. Human enhancements can be used by normal humans to become something closer to these ideal humans. But the specific view of the ideal human one wishes to endorse will necessitate further research to move the debate about the ethics of human enhancement forward, by placing again the notion of the 'human' at the center of the discussion.

Criticizing

Chapter 2: Human enhancement and perfection

Summary

Both bioconservatives and bioliberals should seek to discuss the ideas of human perfection, making explicit their underlying assumptions about what makes a good human life. This is relevant because these basic and often implicit ideas underlying each camp's views inform and influence judgments and choices about human enhancement interventions. Both neglect and polemical but inconsistent use of the complex ideas of perfection have resulted in confusion within the ethical debate about human enhancement interventions. This can be avoided by tackling the notion of perfection directly. In recent debates, bioconservatives have prominently argued against the 'pursuit of perfection' by biotechnological means. In the first part of this chapter, I show that—paradoxically—bioconservatives themselves explicitly embrace specific conceptions of human perfection and perfectionist assumptions about the good human life in order to advocate against the use of enhancement technologies. Yet, I argue that the bioconservative position contains an untenable ambiguity between criticising and endorsing ideas of human perfection. Hence, they still need to clarify their stance on human perfection. In the second part of the paper, I ask whether bioliberals in fact (implicitly) advocate for a particular conception of perfection, or whether they are correct in holding that they do not, maintaining that the concept of perfection is obsolete. I demonstrate that bioliberals also rely on a specific idea of human perfection, based on the idea of autonomy. Hence, their rejection of the relevance of perfection in the debate is unconvincing and should be revised.

Introduction

Bioconservatives such as Michael Sandel, Leon Kass and members of the former US President's Council on Bioethics have expressed skepticism about enhancement technologies, and have argued against bioliberals and transhumanists by resorting to the concept of perfection. They make their 'case against perfection' (Sandel, 2007), warn against 'biotechnology and the pursuit of perfection' (Kass, 2003), or allude to the 'quest of perfection' and its terrible consequences (President's Council, 2003). But what exactly do bioconservatives argue against when they argue against perfection? And how do they make their case?

In the first part of this chapter, I show that—paradoxically—bioconservatives themselves explicitly embrace specific conceptions of human perfection and perfectionist assumptions about the good human life in order to argue against the use of enhancement technologies. Yet, I argue that the bioconservative position contains an untenable ambiguity between criticizing and endorsing ideas of human perfection. In order to resolve this ambiguity, bioconservatives need of clarifying their stance on human perfection.

In the second part of the chapter, I ask whether bioliberals in fact (implicitly) advocate for a particular conception of perfection, or whether they are correct in maintaining that they do not in fact argue for perfection and that the discussion of perfection is obsolete anyway. I show that bioliberals also rely on a specific idea of human perfection based on the idea of autonomy. Hence, their rejection of the relevance of perfection to the debate is unconvincing, and ought to be revised.

Clearly, the complex and abstract concept of perfection has not yet received its due attention in the current debate. If the notion of perfection is used at all, a contentious tone often obscures the debate. Without committing to any specific idea of perfection, I generically understand perfection as stipulating a sufficiently determinate set of ideal human properties that allows for evaluations of enhancement interventions (Hyde, 2010; Passmore, 2000; Foss, 1946). Different ideas of perfection may vary in content, and these differences need to be made explicit. Nevertheless, they serve the same function: making judgments and informing choices. Ideas of human perfection are thus substantially different yet functionally similar. In this study, I want to stimulate an outspoken and constructive debate about the use, role and content of the notion of perfection in the current ethical debate about human enhancement.

Bioconservatives and perfection

Many participants in the moral debate about human enhancement evaluate enhancements along the established and morally salient parameters of safety (Fukuyama, 2002; McKibben, 2003; Mehlman, 2009; Annas et al., 2002), autonomy (Agar, 2004; Buchanan et al., 2000), and justice (Buchanan et al., 2000; Caplan 2009; Habermas, 2003). Although these moral dimensions are also important for Sandel, Kass, members of the President's Council and other bioconservatives (Rubin, 2004; Cohen, 2003), these thinkers agree that these parameters alone do not get to the core of the issue (President's Council, 2003), and ultimately fail to allow for a comprehensive evaluation of a given enhancement practice (Sandel, 2007; Rubin, 2004). In many cases, as bioconservatives argue, these notions are unable to build a case against enhancement (Sandel, 2007; Kass, 2003; President's Council, 2003).

More specifically, bioconservatives oppose the biotechnological vision of *perfection as mastery over human nature*, and being in full control of our constitution and our lives. After illustrating this argument against enhancement, I will show that paradoxically, bioconservatives themselves actually do draw upon a conception of perfection in their case against enhancement.

Skepticism regarding perfection

Bioconservatives often start to build their case against enhancement by expressing skepticism about the 'quest for perfection' that they ascribe to bioliberals, accusing them of being perfectionist in their desire to enhance (Sandel, 2007; Kass, 2003; President's Council, 2003). More specifically, they claim that bioliberals endorse a notion of perfection as mastery over human nature (Sandel, 2007; Kass, 2003; President's Council, 2003), and they fear that pursuing this ideal through biotechnological enhancements would bring forth a type of human who would lose some characteristics essential to leading a good human life.

It might be helpful here to distinguish between the *concept* of perfection and particular *conceptions* of perfection. The concept of perfection is the idea of a sufficiently determinate set of ideal human properties that influences the way a person sees herself and the way she tries to improve. Particular conceptions of perfection are whatever this ideal is about: for some it will mean to have a certain appearance, for others it might mean having certain virtues or acting in a particular way. Conceptions of perfection may differ substantially, but they all function similarly due to the fact that they inform judgments and choices.

In reality, bioconservatives argue against one specific conception of perfection – in which perfect human beings ought to be master of every aspect of their lives through technology. In the current literature, many examples can be found for this claim: Sandel associates enhancement with a form of dominion "that fails to appreciate the gifted character of human powers and achievement" (Sandel, 2007, p. 83). He fears that enhancement will eventually lead to a mastery of the world that would "represent the one-sided triumph of wilfulness over giftedness, of dominion over reverence, of moulding over beholding" (Sandel, 2007). For Sandel, pursuing enhancement technologies is an expression of hubris, and as such, a threat to human virtues such as humility, solidarity and responsibility, which form an integral part of a life lived well.

According to the President's Council and Kass, pursuing perfection and trying to master our human nature would lead to undignified or dehumanised lives. For the President's Council, the use of enhancement technologies might distort what they see as dignity or excellence (President's Council, 2003). For Kass, becoming perfect would have the terrible consequence of becoming posthuman that is, no longer human. Aspiring for such a perfection, which is genuinely non-human, already comes with the risk of dehumanizing (Kass, 2002). In general, Kass is strongly opposed to a view of the perfect life that would include a "painless, suffering-free, and finally, immortal existence" (Kass, 2003).

Other authors like Cohen and Rubin do not explicitly accuse bioliberals of seeking perfection, but resort to a similar argument. For them, the problem of enhancement is that it would alter our state as mortal embodied beings (Rubin, 2004; Cohen, 2003). To surpass our biology with new technology is to go beyond the limits that are given to humans.

These examples illustrate that for bioconservatives, the quest for perfection—understood as mastery over human nature—constitutes a decisive moral concern with human enhancement technologies. This concern goes beyond the moral concerns expressed in the debate in terms of safety, autonomy and justice.

The paradoxical argument against perfection

Paradoxically, bioconservatives themselves rely on a distinct substantive understanding of human perfection while accusing advocates of enhancement of pursuing perfection. The fact that bioconservatives suggest adding another criterion to the established parameters of bioethical reasoning shows their attempt to resort to an ideal of what a human person ought to be. In doing so, they themselves endorse perfectionist ideas, even if they do not explicitly admit to doing so.

For Sandel, the 'perfect' human being should strive to develop the virtues of humility, solidarity, responsibility and gratitude because this is a "proper stance of human beings towards the given world" (Sandel, 2007, p. 9). In his view, pursuing enhancement technologies is morally dangerous because it erodes these fundamental values. Sandel is not opposed to defining a perfectionist ideal along the mentioned virtues for providing a blueprint of how human beings ought to live their lives. What he is opposed to, however, is the biotechnological enhancement of some traits according to an ideal that runs contrary to his own understanding of human perfection.

In this sense, Sandel's *Case against Perfection* actually is not a case against perfection, but a case against a view of perfection that significantly diverges from his own specific understanding of the concept. To have an idea of the perfect human being allows Sandel to stipulate normative limits for the application of biotechnological enhancements on human beings (Baumann, 2010).

Similarly, the President's Council relies on a notion of the perfect human life, defined "not as a life lived with an ageless body or an untroubled soul, but rather a life lived in rhythmed time, mindful of time's limits" (President's Council, 2003). In other words, the perfect human being has to accept, and live within, the limits of the physical body and its death, which give meaning to how a good life should be lived. Without these limits, human life itself becomes meaningless because mortality and the vulnerability (or limits) of the human body enable us to experience real happiness, meaning and agency. It is only within these boundaries that a good human life can flourish. In the view of the President's Council, it is therefore self-defeating to strive to overcome the limits of our bodies. The President's Council's normative framework thus gives a view of human perfection that draws on the human characteristics of embodiment and finality. Like Sandel, the Council relies on a notion of human perfection to evaluate enhancement (Buchanan, 2011), but Sandel's view of perfection centers on human virtues, the Council's view of perfection is about given biological limits (President's Council, 2003).

Both Rubin and Cohen follow the Council's lead. For Rubin, a perfect life ought to be embodied, and this embodiment provides normative limits. In this sense, humans should not try to become posthuman, as it would not be the best human life one can live. Any enhancement that would alter human embodiment is detrimental. For Cohen, humans need to understand what it means to live an excellent life (Cohen, 2003). This human excellence is informed by our biology: the best human life is embodied and limited. The blueprint of human perfection provides a normative framework, which provides in guidance in taking actions and assessing the moral quality of enhancements.

In the spirit of the President's Council, Kass also takes a normative stance by setting forth his own view of human perfection after accusing his opponents of pursuing perfection. For Kass, reasonable views of human perfection should be informed by the fact that humans are mortal, finite and embodied. Therefore, striving for mastery over human nature is detrimental to the pursuit of a good life: embodiment, finitude and mortality of human beings are seen as necessary preconditions of a good human life. If certain limitations of human beings, such as death and embodiment, can be overcome by biotechnological means, or if certain aspects can be controlled, Kass argues that we will no longer be able to experience real happiness, our lives will become meaningless, and our agency will be undermined (Kass, 2002).

In summary, I have shown first that some bioconservatives believe that concerns about enhancement are not adequately captured when only using established bioethical principles such as safety, autonomy and justice. Second, these bioconservatives criticise their bioliberal opponents for defending a misguided conception of human perfection, that is, perfection as mastery over human nature. These differences between bioconservatives and bioliberals seem to mirror the competing fundamental views about the kind of being that seeks perfection. On the bioconservative account, humans are understood as essentially limited beings with an inherent nature and corresponding virtues, while for bioliberals, humans are essentially autonomous and able to alter their very nature. Finally, while bioconservatives argue against perfection, they themselves endorse—at least implicitly—a notion of perfection or perfectionist arguments. Ideas about human perfection rightly understood are, hence, also from a bioconservative's point of view part of the evaluation of enhancement technologies. This, however, gives rise to an untenable ambiguity within the bioconservative position: it is not convincing to blame the opponent for pursuing perfection when one's own view equally relies on an idea of perfection (at least implicitly). Bioconservatives need to clarify the content of their own views about human perfection in order to foster an adequate debate with bioliberals.

Bioliberals and perfection

How do bioliberals react to the bioconservative's reproach? In general, they tend to deny the validity of their criticism altogether by arguing that the notion of perfection is a notoriously vague and potentially dangerous concept that distorts rational discussions and is neither helpful nor necessary to address either the ethical questions raised by the new possibilities of biotechnology or to determine a goal of enhancement-technologies (Caplan and Elliott, 2004).

More specifically, two arguments can be identified in the writings of bioliberals. First, bioliberals deny that the notion of perfection has any relevance for the enhancement debate, since enhancement is only about the *improvement* of human functioning. They thus argue that any discussions about human enhancement that conceive of it as the enterprise of perfection already miss the target at a conceptual level: "The pursuit of biomedical enhancement is not the pursuit of perfection; it is the pursuit of improvement" (Buchanan, 2011). Improvement here stands for the betterment of what went before (Harris, 2007), referring to a former state and not towards an ideal of how an excellent human life should look (understood in this study as the generic idea of perfection).

Second, bioliberals argue against resorting to the concept of perfection on normative grounds: More importantly, since there is no shared understanding of perfection anyway, it would be inappropriate in a liberal society to impose one view of perfection on individuals. The bioconservatives' method of using a concept of perfection to give normative guidelines thus fails to convince the bioliberals. For bioliberals, "there is no such thing as perfection," whether in reality or even as an ideal (Brock, 2009). And even if there were, such an ideal should not be imposed on others in a liberal society. For example, Harris has argued that Sandel's argument is about promoting a Sandelian worldview and outlawing others. Harris fears that such an attitude would lead some to dictate to him what he is allowed to do with enhancement technologies. In other words, he does not want anyone else imposing their ideas of the perfect life, as he wishes to be the only master of his destiny (Harris, 2007). This view gives voice to an important liberal strand in bioethics, according to which we should not invoke any considerations that derive from specific conceptions of human perfection (or, for example, about human nature, the good or the dignified life), because they are based upon controversial assumptions that are not necessarily shared by, and may hence run contrary to, the preferences of the actual members of liberal and pluralistic societies.

In summary, bioliberals argue that we should not draw on the concept of perfection in order to argue for or against certain uses of biotechnology to enhance human beings because enhancement is not about perfection at all (conceptual claim), and because "there is no such thing as perfection, not least because there is no agreed or even agreeable account of what human perfection might consist in" (normative claim) (Harris, 2007).

Can bioliberals do without perfection?

Is the bioliberals' rejection of the bioconservatives' charges convincing? Can and should they argue about enhancement without resorting to human perfection? In this section, I argue for the important role of ideas of human perfection in the debate about human enhancement. I contend that there are good reasons to clarify the notion of perfection as it relates to human enhancement, and that these notions are valuable in the debate about human enhancement, even for bioliberals.

First, the argument that perfection is often used in a vague way and as a rhetorical device by bioconservatives (Caplan, 2009) does not justify the conclusion that this concept should be abandoned altogether. Instead, this claim demands a clarification of the concept and its role rather than an outright rejection. Looking at the different ways in which the concept of perfection is used in the debate, and clearly distinguishing them according to their specific content as well as to their specific function might help promote more productive discussions about human enhancement, particularly with regard to the often-implicit goals of applying technology on humans.

Second, bioliberals claim that improvements can and should be understood only in relation to former states (Harris, 2007) (i.e. how things were before), not by resorting to an ideal state of perfection. However, this approach yields certain problems. On the one hand, talking about 'imperfections' (Harris, 2007) necessarily leads back to some idea of perfection after all (Walker, 2002). On the other hand, arguing about whether something is an improvement only in comparison to former states runs the risk of neglecting certain unwelcome long-term effects. Does a sequence of improvements over former states necessarily lead to an end state that is desirable—whether morally or otherwise? In this regard, the concept of perfection appears to be helpful, perhaps even indispensible, because it calls for an frank discussion of the questions: *what is the end state towards which we are heading when we are improving ourselves and which underlying assumptions justify to identify some change as an improvement?* One possible reason to include discussions about perfection in the debate about enhancement is, then, that a certain shortsightedness, implicit in the liberal approach, can be avoided.

Third, it seems that bioliberals also implicitly endorse a certain idea of perfection while remaining officially neutral about it. If they are only concerned with moral constraints on enhancements, and leave it up to individuals to decide whether to engage in enhancements, this indeed presupposes some ideal of (individual) mastery over human nature (i.e. regarded as distorted by bioconservatives). This is based on a strong understanding of an individual's *autonomy*, a concept that figures prominently within the bioliberal ideal of a human life.

This leads to the following question: if it were not for an explicit discussion of ideals of perfection, would the certain ways in which societal ideals of perfection influence individuals escape bioliberals' attention? By stressing individual autonomy, they neglect to consider the influence that ideals of perfection—such as, for instance, cultural images of the 'perfect body'—have on individual decisions. Making explicit the underlying ideals of perfection that are present in our societies and that certainly have some degree of influence our views about enhancements can help to account for this important social dimension in the debate.

Even though the concept of perfection is often used in a vague or contentious way, and even if specific views about human perfection are controversial, it should still play a significant role in the debate about human enhancement. All those engaged in the debate, particularly the bioliberals, should not shy away from discussing the often implicit and complex ideas of perfection that guide human judgments and choices. After all, bioliberals—despite their claims—rely on a specific ideal of humans, in that case as autonomous beings. This should be understood as a specific idea of perfection, and thus requires a thorough confrontation with competing assumptions about human perfection.

Conclusion

While bioconservatives accuse bioliberal proponents of enhancement technologies to strive for perfection, understood as mastery over human nature, paradoxically, it is mainly the bioconservatives themselves that rely—even if often only implicitly—on specific conceptions of perfection in terms of what constitutes an ideal human life, when it comes to evaluating the use of enhancement interventions. Accordingly, bioconservatives should not argue against the pursuit of perfection altogether, but should instead focus their critique on the specific understanding of perfection they claim to find endorsed in bioliberal views. In doing so, they will have to take into account the fact that many bioliberals deny defending any concept of perfection and prefer only to speak cautiously of means of improvement.

The bioconservatives' position endorses two main claims about perfection. First, they believe it is methodologically beneficial to discuss human perfection—in addition to established moral parameters like safety, autonomy and justice—as an important normative standard to evaluate technologies of enhancement. Second, reflections on the content of human perfection, rightly understood, justify a restrictive attitude towards biotechnological enhancements. Bioliberals, with their erroneous view of perfection as mastery over human nature, falsely argue for a permissive stance towards enhancements, as the bioconservatives argue.

Bioliberals reject both these views because to them notions of human perfection seem irrelevant for assessing the moral quality of enhancements. At best, an understanding of (relative) improvement will suffice. In any case, the established forms of moral reasoning can do the job of a moral assessment. And since bioliberals claim not to endorse any view of human perfection at all, the criticism from their bioconservative opponents does not bother them.

However, as I have attempted to show, raising the question about enhancement and perfection brings up the crucial question of the ultimate goals of biotechnological interventions, which might depend on an idea of what it means to live a good or a perfect human life, an idea that may provide guidance and allow moral evaluations of specific enhancement interventions. Even if one is not convinced by the bioconservatives' specific suggestion about the content of human perfection, it is certainly to their benefit to have brought up the case. It is doubtful whether anyone—including bioliberals—can reject a discussion of the idea of human perfection without neglecting an important part of the moral assessment of enhancement. For a comprehensive moral evaluation of enhancements, a frank debate about human perfection and the essential elements of a good human life is urgently needed. Failing to take into consideration the often-implicit assumptions about human ideals that inevitably influence and direct choices about enhancement interventions will cause us to miss important moral insights into these issues.

Conceptualizing

Chapter 3: Evaluating human enhancements: the importance of ideals

Summary

Is it necessary to have an ideal of perfection in mind to identify and evaluate true biotechnological human 'enhancements'? To answer this question, I call upon the distinction between ideal and non-ideal theory, found in the political philosophy debate on theories of justice. This debate brings together the distinctive views about whether one needs an idea of a perfectly just society, or not, when it comes to assessing the current situation and recommending steps towards increasing justice. In this chapter, I argue that evaluating human enhancements from a non-ideal perspective has some serious shortcomings, which can be avoided by endorsing an ideal approach. My argument begins with a definition of human enhancement as improvement, which can be understood in two ways: the first approach is backward-looking and assesses improvements with regard to a *status quo ante*, and the second, forward-looking approach evaluates improvements with regard to their proximity to a goal or according to an ideal. After outlining the limitations of an exclusively backward-looking view (non-ideal theory), I answer possible objections to the forward-looking view (ideal theory). Ultimately, I argue that the human enhancement debate lacks some important moral insights if a forward-looking view of improvement is not taken into account.

Ideal and non-ideal views about human enhancement

When it comes to making moral judgments about human enhancement interventions, one methodologically important question is whether we need ideals or assumptions about perfection in order to make these judgments, or whether it is sufficient to assess the enhanced state of a person in purely relative terms by comparing it with the former *status quo ante*. In this chapter, I argue that evaluating human enhancement exclusively from a non-ideal perspective that does not consider any specific ideals has some serious shortcomings, which can only be avoided by complementing this perspective with the ideal approach. I understand "perfection" as a set of ideal human characteristics or traits that allow for evaluations of enhancement interventions. Different ideas of perfection may vary in terms of content, and these differences need to be made explicit; nevertheless, they serve the same function of making judgments and informing choices. Ideas of human perfection are thus substantially different but functionally similar (Roduit et al., 2013).

The argument unfolds in three steps: (1) Starting from a definition of human enhancement as improvement, I distinguish two ways of understanding improvement. The first approach is backward-looking and assesses improvements with regard to a *status quo ante*. The second, forward-looking approach evaluates improvements with regard to their proximity to a goal or according to an ideal. (2) The assessment of relative improvements with the help of an exclusively backward-looking, comparative or non-ideal approach results some substantial limitations, as these approaches are short-sighted and incapable of providing a clear objective for enhancements and it only appear to be free from substantial assumptions about ideals. (3) On the other hand, there are also warranted objections against any ideal approach that prescribes an ideal of human perfection, as these definitions may be insufficient, unnecessary, intolerant or inflexible. Yet, these difficulties can be moderated when the relationship between an ideal and a non-ideal approach is not understood as competitive, but as complementary.

The overall aim of the argument is to make a methodological point, not to designate or defend what such an ideal would look like. This chapter emphasizes that a parallel can be drawn between the political philosophy debate and the enhancement debate. It then goes on to analyze what consequences this holds for the debate on human enhancement.

Human enhancement as improvement

Various definitions of human enhancement have been suggested in the debate. An important definition contrasts enhancements with therapeutic interventions and stipulates that the former are outside of the scope of the latter. In this sense, enhancements are interventions going beyond therapy (Allhoff et al., 2010; Allhoff and Steinberg 2011; President's Council, 2003; Daniels, 2000; Juengst, 1998). Another noteworthy definition states that HEs are quantitative changes that add to an existing state or set of abilities (Chadwick, 2008, 2011; Naam, 2005). The third important definition refers to human enhancement as qualitative change. In this view, the evaluative notion of *better* is implied (Buchanan, 2011a; Buchanan, 2011b; Harris, 2007;

Pence, 2012; Sandel, 2007; Savulescu, 2006). In comparison to the other definitions, understanding human enhancement as a kind of improvement both provides a generally positive characterization and introduces a qualitative element. Because this definition conceives of enhancements as biotechnological *improvements* and adds a qualitative aspect, is a particularly plausible definition, and will not further discuss the first two definitions.[6] In this chapter, I therefore understand human enhancement as an intervention in the human body by biotechnological means to *improve* – from some perspective, in some regard, and in a certain domain – the condition of an individual (Buchanan, 2011a).

Two ways of looking at improvement: backward and forward

But what counts as an improvement? There are two different ways of evaluating improvement: on the one hand, improvement can be evaluated in reference to a former state (*status quo ante*). Enhancement, in this *backward-looking* view, is "by definition an improvement on what went on before" (Harris, 2007, p. 9). The backward-looking view of improvement attempts to seize the opportunities provided by biotechnological interventions to make things relatively better. This view does not require having a specific ideal or end point in mind in order to identify an improvement, as reference to a former state is already sufficient (Buchanan, 2011a; Harris, 2009). In this view, enhancements can be understood as "[an] improvement relative to the present state of affairs, [which] aims to remove or repair human flaws without prejudgment about what is 'ideal' or perfectly human" (Mahootian, 2012, p. 143). By framing the concept of human enhancement as something not intrinsically linked to any particular ideal, this definition is more conducive to the needs of a liberal, pluralistic society.

However, this view invites critical questions regarding the ends of enhancement (Walker, 2002; Shuman, 1999; Hanson, 1999; Keenan, 1999; Mckenny, 1999; Baertschi, 2011). When attempting to improve some traits, does one aim to make them better because they are imperfections,

---

6 For articles dealing specifically with definitions of human enhancement, see: Menuz, 2011; Savulescu, 2006; Chadwick, 2008; Resnick, 2000.

or because they are not aligned with certain ideals? Does the concept of imperfection itself assume some conception of an ideal or even perfection? According to the critics of the backward looking view, an improvement might be evaluated with regard to its goal (Walker, 2002; Keenan, 1999) or a certain ideal (Baertschi, 2009; Grunwald, 2009). Such a forward-looking view is a major alternative to understanding improvements only in relation to the *status quo ante* because the proximity of an enhancement intervention to an ideal is taken into account.

We can hence distinguish between two ways of understanding enhancements as improvements: as departing from a *status quo ante* on the one hand, or as approaching a guiding ideal on the other. In the latter case, it is the ideal that influences the direction of the human enhancement project, not the identification of flaws in existing circumstances, as is true in the former case. Baertschi illustrates the forward-looking, ideal approach as follows:

> [N]either our desire for enhancement nor our concerns about personal identity can be properly understood without referring to an explicit or implicit ideal: *the ideal of the person we want to be*. This ideal is an essential part of our conception of the good life, because a good life is a life we want to live, as the person we want to be. (*emphasis mine*, Baertschi, 2009, p. 39)

These two distinct ways of evaluating enhancement as improvement have heretofore been undifferentiated in the literature and have not been paid due attention.

Ideal and non-ideal approaches to evaluate human enhancement

When distinguishing between a forward-looking and a backward-looking understanding of enhancement as improvement, the question arises as to whether these two positions are antagonistic or complementary. This discussion shares some interesting similarities with the political philosophy debate about non-ideal and ideal theories of justice. Do we need to have a substantial ideal of a perfectly just society in order to identify adequate improvements of the current state, or will a non-ideal theory that is able to identify relative improvements while remaining agnostic about potentially perfect end-states suffice?

In political philosophy, Sen recently scrutinized these two approaches (Sen, 2006, 2009). He believes a theory to be ideal when it is not comparative, but absolute (Valentini, 2011). A theory can be said to be absolute when it focuses on identifying perfection, instead of comparing two alternatives that are less than perfect (Sen, 2009). Here is a working definition of a non-ideal and an ideal approach to justice in society:

- X is better than Y. Under the comparative approach, one society (X) is judged to be relatively better (i.e. more just) than another (Y). This is the non-ideal approach.
- X is perfect, given that a specific condition or conditions are met. This transcendental or ideal approach outlines a perfectly just society (X). Under this approach, existing states are judged according to how close they come to the ideal (Valentini, 2010).

Valentini has outlined three possible interpretations of the ideal vs. non-ideal theory in political theory: 1) *full compliance* vs. *partial compliance* theory; 2) *utopian* vs. *realistic* theory; and 3) *end-state* vs. *transitional* theory (Valentini, 2010). For our purposes here, I will focus on the third interpretation (*end-state* vs. *transitional* theory), because the end-state refers to an ideal or an end, while the transitional aims at comparing two alternatives in order to find the better option.[7] The transitional approach does not presuppose an ideal end-state, but assumes a set of dimensions according to which relative progress can be assessed. Because of this, it is non-ideal because the current states in and of themselves have not yet reached an (indeterminate) ideal state during the ongoing transitional period.

A comparative, transitional approach identifies one of the two states that are compared with one another as 'better' or 'worse', without either an absolute positive ideal and without an idea of imperfection. It is simply a relative comparison between two states.

Applying these distinctions from political philosophy to the debate about human enhancement provides ways of looking at enhancement from different perspectives:

- The backward-looking, comparative or non-ideal approach asks, "Would it be better if human had X?" Here, two states are compared with regard to a specific dimension of human existence, and the better state is

---

7   For more details on the other view, see Valentini, 2012.

identified without resorting to a fully fleshed-out, complex ideal of human existence.
- The forward-looking, end-state, or ideal approach asks, "What would be a perfect or ideal human being?" Here, assumptions about an ideal are the precondition for evaluating states regarding to their proximity to this ideal.

So, to evaluate enhancing interventions, is it sufficient to employ comparative criteria (i.e. non-ideal), or must we also have an ideal standard of a perfect end-point in mind? A purely backward-looking view ignores any ideals, as these are considered unnecessary and problematic for the various reasons. This view, as I will show in the next section, has substantial limitations. Ultimately, I argue that evaluating human enhancements solely from a backward-looking perspective – without taking into consideration a forward-looking approach – has some serious shortcomings. The suggested analogy between the debate in political theory and the use of it in bioethics generates, as we will see, some interesting critique on the backward-looking and forward-looking ways of evaluating enhancement. In other words, the *arguments* used in the political debate between ideal/nonideal theory can enlighten the debate regarding the ethics of human enhancement.

Limitations of a non-ideal approach

Following a non-ideal approach is subject to several criticisms, including a lack of objective and general shortsightedness, as well as having the mere appearance of neutrality.

Lack of an objective and shortsightedness

A first problem with a non-ideal approach is that it lacks a specific objective and thus is inevitably shortsighted. When speaking of a non-ideal theory of justice, Rawls mentions that "until the ideal is identified [...] non-ideal theory lacks an objective, an aim, by reference to which its queries can be answered" (Rawls cited in Valentini, 2012, p. 660). In the following example, Robeyns' point fits well here and is worth quoting at length:

> Suppose that we can represent the degree of justice of a certain situation with a number, on a scale where 100 represents the fully just social state. The initial social state A has a justice value of 50. From A, we can move to either B or S, with B corresponding to a justice value of 70 and S of 55. If we are in A, and only compare

B and S, then the conclusion is easy: we have to take action so that we end up in social state B. But our possibilities for further action are not independent of this first choice. Suppose that in the best-case scenario we can move from B to C, with C having a justice value of 80. From S, however, we will be able to move to T where we can realise a justice value of 95. We cannot move from B to T. It then becomes clear that in order to make a reasonable decision between B and S, we need to know the 'paths of change' that B and S are on, and those paths are directing us towards an ideal, that is, a transcendental theory (Robeyns, 2012, p. 160–1).

Therefore, in order to be able to evaluate an improvement, it is essential to clarify the ideal we are aiming to achieve (Simmons, 2010). This is where ideal theory becomes helpful, as it "dictates the objective…[and] the route to that objective from whatever imperfectly just condition a society happens to occupy" (Simmons, 2010, p. 12).

An analogous argument to the one made in the theory of justice can also be put forth in the debate about human enhancement. Indeed, it seems difficult to know whether one is on the right paths, if the end-state for an improvement remains unspecified. For this reason, when considering enhancement, an important question is "Towards what do we improve?" (Walker, 2002). This question cannot be avoided or answered with a purely gradual approach that allows for relative improvements in certain dimensions of human existence without first defining an ideal. After all, without an ideal, we cannot identify the moment when further relative changes that were considered improvements stop being mere improvements and become the ideal. If qualitative judgments about improvements are based on measurable quantitative changes (as it is the case in biotechnological enhancement interventions), it is clear that relative increases ad infinitum will most likely not bring about further increases in quality ad infinitum. Relative improvements would only take place up to a certain upper limit for traits like intelligence, empathy, and physical strength, and so on. There is a strong need to spell out an ideal in order to know at which point further changes will stop being true improvements.

If no comprehensive objective of enhancement interventions is defined, an evaluation of enhancements runs the risk of being shortsighted, meaning that while judgments would be able to identify the immediate advantages or improvements of an enhancement, but these evaluations would fail to take into account long-term consequences of the enhancement, which may indeed be less advantageous and perhaps even outweigh the short-term gains. It seems plausible to assume that a more specific and clearly stated goal or end-point of enhancement interventions could help avoid such shortsightedness.

For example, if one considers becoming taller only in comparison to a given former state without regard to an end-point, the enhancement is only a quantitative change. It does not follow that such a change is an improvement. On the other hand, if one takes an ideal approach, the goal and ends of human enhancement become central factors guiding the evaluation. If one undergoes a height-adding enhancement, for example, out of a desire to become a better basketball player, the enhancement can be considered a qualitative change, once the intervention is positively evaluated in reference to the ideal of what is required to be a better basketball player.

The appearance of neutrality

Non-idealists claim to remain neutral, silent or agnostic on the question of human perfection or the ultimate goals of improving human beings. Instead, they want to leave it up to the individual to choose their goals and forge their life plans according to their individual preferences. In this approach, non-ideal approaches conform with the prevailing view that in liberal societies, choices about individual well-being and individual goals are not a matter to be settled in the general public or be prescribed to individuals.

Whether it is possible to uphold this appearance of neutrality when it comes to ideals is questionable, since it seems that in some cases the pluralist, liberal view defended by the non-idealists contains in itself a substantial ideal of human well-being. Being able to use available biotechnological means to pursue human enhancement according to one's own individual preferences is based on the idea of the human self as an autonomous agent who bears exclusive responsibility for one's own life (Roduit et al., 2013). In this context, emerging biotechnological interventions become helpful to the individual who wants to – and should be able to – exercise mastery over her own life.

The apparently neutral, non-ideal view seems to endorse specifically Western, contemporary societal ideals about autonomous agents, and fails to acknowledge the impact that societal ideas have on the formation of individual preferences. Instead of being truly neutral, the call for a plurality of preferences and range of ideals on perfection characterized by the non-ideal approach is in fact the result of viewing human beings as autonomous agents. For instance, if an individual freely chooses to pursue some sort of cosmetic enhancement, claiming to be free to control his life, he should be aware that different ideals of beauty have certainly influenced his so-called autonomous decision. Making this decision is actually not as autonomous as it might first appear.

As a consequence, proponents of the non-ideal, backward-looking approach are unable to claim neutrality. This becomes obvious in the ethical evaluations that are made on these grounds. For example, it has been argued that we have a moral obligation as humans to enhance our children by utilizing technological advancements that would allow them to overcome imperfections and to fully realize what counts as valuable (Harris, 2007). While at first purporting to be neutral, in order to allow individuals to make up their own mind about perfection, this view may, however, end up claiming that such or such enhancement is mandatory. Therefore ending not being neutral at all. Imposing a given enhancement on someone is totally contrary to allowing individuals to make autonomous choices. Here, certain perfectionist assumptions are implicitly used, and the debate of human enhancement stands to gain clarity by making these assumptions explicit.

Perfectionist elements are also implicit in the non-ideal approach towards evaluating enhancements. It is advisable to tackle the question of ideals directly, since even accounts that claim not to refer to any ideals do implicitly contain them.

Defending an ideal approach

In this methodological chapter, I do not aim to spell out a substantial account of an ideal that could or should be used to guide evaluations of human enhancement interventions. Instead, I argue in favor of allowing for substantial reasoning about human ideals in order to provide a comprehensive assessment of enhancement interventions in combination with, and as a complement to non-ideal considerations. In the following section, I defend the need for an ideal approach by addressing and rejecting possible objections to the necessity of providing a definitive account of human perfection.

Insufficient

Two prominent objections have been raised to ideal theory in political philosophy: the *sufficiency* and *necessity* arguments (Sen, 2006). The former claims that an ideal is incapable of deciding which of two non-ideal alternatives is preferable. Just having an ideal is not enough to judge which of two possibly very different states is closer to it than the other. Accordingly, perfection does not help to make "decisive comparisons among imperfect alternatives" (Sen, 2009, p. 97). Only a comparative approach, non-ideal one that weighs two options directly against one another can do this.

It has been argued, however, that identifying an ideal is helpful when it comes to comparing non-ideal alternatives. Having an ideal reference point is informative because as it defines the end point at which attempts to further improve could be ceased. Its directive power is also not to be underestimated: knowing what would be the ideal or end-state clearly determines the general direction in which improvements are undertaken. Possible cases of conflict or indeterminateness notwithstanding, knowing one's goals is essential for taking meaningful and reasonable action.

With regard to the debate about human enhancement, one objection is that it is not helpful to know about a perfect human being when it comes to making a choice between two alternative options that both fulfill some criteria of a perfect being, but not others. Non-idealists further claim that knowing about perfection does not help identify those imperfections that are worth being improved (Brock, 2009; Harris, 2007).

However, idealists do not have to claim that an ideal *alone* is sufficient to evaluate enhancements; other normative concepts such as safety, justice, and autonomy also play a role. Yet in many cases, the idealist would argue that it is helpful and necessary to resort to an ideal, which serves to complement these ethical tools in the evaluation of human enhancement. As I have shown, both sides seem to have a point here, but when it comes to conducting a comprehensive evaluation of enhancement interventions, both the comparative assessment and more general assessments about the overall direction of one's ambitions play an important role.

Unnecessary

The second argument against ideal theory claims that an ideal is not only insufficient, but also unhelpful and unnecessary for evaluating or recommending enhancement options. What matters is comparing the alternatives and options themselves, for which the idea of a perfect but remote endpoint is simply of no relevance. Looking at obvious imperfections, diseases, and disabilities, and contrasting them with relatively better alternative states, would accomplish the aim of providing decisive evaluations.

Sen suggests an analogy: if one had to compare a painting by Picasso and a painting by van Gogh, it would not help to know that the da Vinci's Mona Lisa is the best painting in the world (Sen, 2009). The heterogeneous styles, ambitions and techniques involved make it impossible to resort to a single standard of perfection. In many cases, Sen argues, the answer to the question of proximity to an ideal – if there is even an ideal in the first place – is superfluous.

However, Sen's argument by analogy is misleading here. To know that the Mona Lisa is the best painting would indeed help to evaluate other paintings insofar as this knowledge would provide additional criteria for making a judgment of perfection, like creativity, use of color, use of shade and other techniques. These criteria can then be transferred to evaluate other paintings.

Clearly, an assessment guided by an ideal would not have to be the only evaluative tool, but one that would play an essential role in the process of evaluation.

Similarly, in the debate about human enhancement, proponents of the backward-looking approach claim that knowledge about ideals is not necessary to identify human imperfections and recommend corresponding interventions to overcome these limitations (Harris, 2007). Yet, without having an ideal in mind, which would justify why certain human imperfections may not turn out to be desirable, all things considered, and even apparently uncontroversial judgments about imperfections (and corresponding advice to enhance) would be uninformed. For a full assessment of enhancement options, it is necessary to think about the ultimate ends an enhancement is aimed at. Otherwise, one would run the risk of going astray with regard to the general direction of development.

Pluralism and sufficiency

In a world that cherishes a plurality of values and different lifestyles, a single comprehensive set of ideals is met with skepticism. At least in today's diverse Western societies, there seems to be a general consensus that there can be more than one valid ideal of a good human life. Similarly with regard to the distinction between ideal and non-ideal theory in political philosophy, one could argue that there are simply several equally good paths towards creating a just society. So, how might knowledge and understanding about an ideal help in justifying evaluations of an improvement in a widely pluralistic society?

Here, it is important not to overestimate the power of an ideal approach. It is not about answering all questions only in reference to an ideal, but this ideal should play an important role – even when a diversity of values is legitimate and desired – in identifying the central or core values that are necessary for living a good human life. Defining the basic needs, central human capabilities or essential conditions for human well-being help stipulate a vision of an ideal life that sets at least a minimum threshold.

It may appear questionable whether stipulating a minimum threshold can count as an ideal or a state of perfection, but in the view presented here it makes perfect sense to argue that having or being *enough* can well be a viable factor for determining human perfection. In this reasoning, it should become obvious that an ideal does not have to be an intolerant, exclusive, or coercive view about a specific way of living a good life, e.g. Harris' fear that Sandel would impose his particular world view on him (Harris, 2007). One such ideal, however, may serve the basic function of helping to determine the basic prerequisites needed for being free to live a good life. This ideal could also be broad enough to allow for multiple individual choices and preferences.

Contrary to Harris' view, there is a continuous need for analyzing and evaluating the best arguments about different conceptions of the good human life. While there might be a plurality of views, it remains important to assess those views according to their underlying assumptions pertaining to what is of ultimate importance, or what should count as perfect. This is especially important in a diverse society with many competing worldviews.

Inflexible

A remaining issue for the ideal theory approach is that it might seem inflexible, narrow-minded and biased towards the status quo (Valentini, 2011), and this objection applies both to the debate about justice and to the debate about enhancement. Once an ideal is established, we become reluctant to modify or revise it: "We remain trapped, so to speak, in the realm of perfection" (Valentini, 2011, p. 302). But contrary to the non-ideal view, the solution is not to get rid of perfectionist notions altogether, but to evaluate different conceptions and to make sure they remain open to revision as new possibilities for improvement arise.

This, of course, is a difficult challenge that needs to be addressed on a societal level. Avoiding set, narrow views is generally preferable, and the ends of human enhancement interventions should not be seen as static and eternally fixed. Even the idea of human perfection may change according to the circumstances. The presumed endpoint that is taken to evaluate and assess human enhancement interventions should hence be understood as an "end in view" that directs our efforts for improvements without claiming absoluteness or being the last word on what makes a good human life.

Conclusion

The ethical debate about human enhancement focuses on the question what can be considered a permissible and recommendable improvement from a moral point of view. In this chapter, I discussed the role that ideas of perfection play in this debate. I have contrasted competing views: the non-ideal, comparative, backward-looking approach, and the ideal, goal-oriented, forward-looking approach. I have argued for a comprehensive approach, in which both approaches mentioned above play an essential role. Since the ideal approach in particular faces substantial criticism, I have devoted a good deal of attention to the challenge of showing the importance of ideals of human perfection in the debate, as an exclusively non-ideal approach would have serious shortcomings. I have not attempted to define a precise ideal that would be worth pursuing or that would enable justified moral evaluations of enhancement interventions. This remains the task of the next two chapters.

Constructing

Chapter 4: Rejecting problematic conceptions of perfection

Summary

Whatever ethical stance one takes in the debate regarding the ethics of human enhancement, one or more reference points are required in order to assess the morality of the practice. Some have suggested looking at the bioethical notions of safety, justice, and/or autonomy to find such reference points, while others introduce some perfectionist assumptions into the debate, arguing that these bioethical notions are limited with respect to assessing the morality of human enhancement, and have instead turned to human nature, human authenticity, or human dignity as reference points. In Chapters 4 and 5, I ask which perfectionist assumptions should be used in this debate. Chapter 4 seeks to make explicit the perfectionist assumptions found in the debate, and eliminate those that are problematic. Chapter 5 will outline a solution.

Introduction

Whatever stance one takes in the debate regarding the ethics of human enhancement, reference points are required to assess its morality. Some have suggested looking at the frequently used bioethical notions of safety (Annas et al., 2002; Fukuyama, 2003; McKibben, 2004; Mehlman, 2009), justice (Buchanan et al., 2001; Habermas, 2003; Schermer, 2008; Mitchell, 2009; Zylinska, 2010), and/or autonomy (Buchanan et al., 2001; Agar, 2004; Schaefer et al., 2014) to find such reference points. Others, arguing that these bioethical notions are limited when it comes to assessing the morality of human enhancement (Roduit et al., 2013), have turned to human nature (Heilinger, 2010; Heilinger, 2014), human authenticity (DeGrazia, 2000; DeGrazia, 2005; Parens, 2005; Levy, 2011; Erler, 2012), or human dignity (Kass, 2003; President's Council, 2003; Kass, 2004) as reference points.

As suggested elsewhere (Roduit et al., 2014), one could also find such a reference point by looking at the 'ideal' human. After all, the goal of human enhancement is not necessarily to become an authentic or a normal human but to become an ideal or even a perfect one. Accordingly, this vision of human perfection could serve as a reference point to guide enhancements. Although it is not possible for all people to agree on what exactly an ideal human would look like in all circumstances or at all times, we can nonetheless hypothesize what some characteristics of an ideal human would be. I refer to these characteristics here as perfectionist assumptions of what it means to live a good human life. In other words, to be considered an ideal human, one would have a "set of ideal human properties that allows for evaluations of enhancement interventions" (Roduit et al., 2013). As is written elsewhere: "Different ideas of perfection may vary in content, and these differences need to be made explicit; nevertheless, they serve the same function of making judgments and informing choices. Ideas of human perfection, hence, are substantially different but functionally similar" (Roduit et al., 2013). These perfectionist assumptions become an additional reference point for evaluating the morality of an enhancing modification. If a given human enhancement moves in the direction of the chosen ideal, it will be seen as morally acceptable, assuming there are no moral concerns with other related issues of justice, safety and autonomy.

In here and the following chapter, I ask what views of human perfection – or which particular human perfectionist assumptions – should be used in the debate regarding human enhancement. This chapter criticizes views that are problematic, while the next takes a positive approach and suggests some perfectionist elements based on Nussbaum's capabilities approach that can lend guidance to the practice of human enhancement.

Ultimately, the aim of this section is twofold. First, it seeks to make explicit the perfectionist assumptions found in this debate and eliminate those that are problematic. Second, it clarifies an element that is often neglected in the debate about human enhancement: Towards what view of the ideal human should human enhancement aim? Here, I outline central capabilities that are essential for an ideal human to possess and can therefore serve as reference points to guide human enhancement. The focus of this section is only on *human* enhancement, not on other types of enhancement. It does not address whether it is morally acceptable to enhance, for example, from an animal to a human, from a human to a posthuman, or from a posthuman to an animal.

Content of Perfection

When examining the various understandings of human perfection, it is useful to have a closer look at the different types of perfectionist assumptions found in the debate. Before proceeding, a terminological clarification is in order: although perfection refers to any objectivist account of well-being in the debate of political philosophy for some, I am using the term here in the sense of what an ideal, perfect human being would be like, that is a "set of ideal human properties that allows for evaluations of enhancement interventions" (Roduit et al., 2013).

We can distinguish between two conceptions of perfection: type-perfection and property-perfection (Walker, 2002). Type-perfection is "the thesis that those individuals who best realize the essential properties of the individual's type or species best exemplify the ideal of perfection" (Walker, 2002). The second concept of property-perfection is the "thesis that those individuals who best realize some property or properties best exemplify the ideal of perfection" (Walker, 2002). In this account, one refuses to take into consideration any sort of type when evaluating or pursuing human enhancement. What matters is the enhancement of different properties, independently of a type. Perfection is thus realized in its relation to a property rather than to a type. This allows a property to be fully perfected without the limitations that might come with a type.

Proponents of type-perfection approve of using human enhancements within the limits of a type, being careful not to enhance too far, which runs the risk of becoming something other than the type given. Proponents of property-perfection will seek to enhance one particular function, without making reference to any given type, even if this means becoming something other than the given type. Walker uses the example of enhancing the intelligence of a monkey to lay clear the distinction between both concepts. For proponents of type-perfection, the intelligence of a monkey can be improved only if a monkey remains a monkey. For property-perfectionists, it is not the monkey but the property of intelligence that should be perfected to its maximum, even if this means that this particular monkey will become something other than a monkey, a post-monkey so to speak (Walker, 2002).

Although the specifics regarding what the ideal human would look like are not always fully fleshed out in the debate, some type-perfectionist assumptions can still be found. These type-perfectionist assumptions can take different forms, as we will later see in more detail. Briefly stated, they can be a set of human virtues (Keenan, 1999; Sandel, 2007), a set of human limitations (Cohen, 2003; President's Council, 2003; Rubin, 2004; Agar,

2010), a set of human characteristics (Parens, 1995; Fukuyama, 2003), or even the ideal of autonomous individuals who master every part of their human nature (Harris, 2007; Brock, 2009). In the debate, if these essential properties can be fulfilled, an individual would then exemplify the ideal of perfection according to its particular type. This view is generally used to argue against human enhancement (Cohen, 2003; Fukuyama, 2003; Kass, 2003; President's Council, 2003; Rubin, 2004; Sandel, 2007), and in some cases against certain types of human enhancement (Parens, 1995; Agar, 2010). Less commonly, it has also been used to argue for human enhancement (Harris, 2007; Brock, 2009).

Property-perfection can take two different forms. It can be understood as a set of defined objective goods, in which, for example, properties such as intelligence, memory or others ought to be maximized independently of a given type. Alternatively, it can be understood as a set of undefined subjective goods, which individuals are free to pursue what seems good as they define for themselves (Walker, 2002). As Walker explains, "One step, and seemingly the first, would be to decide which properties of ourselves we would like to develop, which we would like to see remain unchanged, and which we might hope to eliminate" (Walker, 2002).[8]

However, the property-perfectionist view is flawed as a self-sufficient position because it ultimately falls back on the type-perfectionist view. To speak of property alone is incomplete because a property is always associated with a type. For example, when we speak of the perfect speed, we do so in reference to a certain type, such as an athlete. The property of speed has to be embodied in a type, which could of course take different forms such as a human, a posthuman, another species or even the form of an object. In all of these examples, however, it is embodied in a type. When people defend a property-perfectionist view, arguing that some properties should be enhanced without reference to a type, they overlook the fact that a property

---

8 In addition to these two views, a third one should also be mentioned here: the anti-perfectionist view. This view claims that to speak of perfection at all in this debate is useless because human enhancement is not conceptually related to perfection. However, as it has been shown elsewhere, the anti-perfectionist view has some serious shortcomings and is untenable because it implicitly embraces some perfectionist notions. See Roduit et al., 2013.

cannot exist without a type. What they sometimes do is put forward a post-human type in contrast to a particular human-type. Proponents of this view disagree that a type can be given by nature or by god. However, a type can be socially constructed and democratically agreed upon. Thus, there is no need, at least for now, to determine whether a type is given or constructed. To show that we cannot ignore type-perfection is sufficient here.

Sources of Perfection

Where do these perfectionist assumptions come from? Ideas about perfection in this debate come from two main sources: the subjective view and the objective view. According to the subjective view, one creates her own type-perfection or property-perfection, such as a list of subjective goods. Menuz et al. tend to support this view, arguing that human enhancement has to be defined according to individuals. They assert that "human enhancement is defined a posteriori as any intervention aiming to modify biological and/or psychological features—using NBICs—if, and only if, it actually allows an individual either to reach or to improve what she/he considers her/his personal optimum state" (Menuz et al., 2011). Agar defends this view in *Liberal Eugenics*, where he argues that one is free to enhance as long the enhancement does not result in personal harm (Agar, 2004). However, in *Humanity's End*, he defends a type-perfectionist perspective against some forms of human enhancement (Agar, 2010). Walker also argues for different properties that need to be perfected, but leaves room for individuals to decide whether or not to enhance (Walker, 2002). However, in his book *Happy-People-Pills for All*, he does develop a comprehensive view of the good human life, and thus ends up arguing for an objective list (Walker, 2013).

The subjective approach is deeply connected to the ethics of authenticity, which hold that, above all else, one should be oneself, become oneself and stay true to oneself. Within this view, the individual is free to enhance towards the ideal he has subjectively chosen for himself. The only restriction here, as part of a social contract, is the no harm principle, which would allow people to enhance themselves as long as they are not harming others.

According to the objective view, the common good is either objectively outlined in a list of goods without any reference to a type, or outlined by some type-perfectionists notions. In the debate, there are different ways to undertake this task. Savulescu outlines some non-exhaustive objective goods such as intelligence, memory, self-discipline, impulse control, foresight, patience, humor, sunny temperament, empathy, imagination, sympathy, fairness, and honesty (Savulescu, 2007). Buchanan et al. state that these goods are introduced as general-purpose means, which are properties valuable to anyone regardless of what their view of the good life entails (Buchanan et al., 2001). Baertschi hints towards Sen's capabilities approach for some objective references (Baertschi, 2011), and Hughes suggests looking at the virtues and the capabilities approaches of Sen and Nussbaum (Hughes, 2011). Kass resorts to human dignity (Kass, 2003), while Sandel utilizes a list of virtues (Sandel, 2007). Finally, Harris uses the value of mastery over every aspect of human nature (Harris, 2007). Should we therefore endorse an objective or subjective approach?

Problems with a subjective approach

The main problem with the subjective approach is that it could lead to mere relativism. To remedy this problem, a given society could agree – as part of a social contract of sorts – to use the non-harm principle as a safeguard. Individuals would be free to enhance as long as they do not hurt others, which would enable individuals to be as autonomous as possible.

This subjective stance, however, does not stand on its own, either. Outlining a particular conception of the non-harm principle introduces an objective component into the debate, and the agreed-upon principle becomes an objective standard in and of itself. Moreover, the non-harm principle would also have to take into consideration problems or injuries caused by the enhancement that stem from social pressure (Brownsword, 2012; Brownsword, 2013), lack of education, the inability to relate to the surrounding world, social disruption, and other causes. As we will see later, these potential problems relate to a set of capabilities. Taken together, one could therefore establish a list of some objective capabilities that humans should have in order to live a good human life. Thus, the subjective approach, even with an agreed-upon non-harm principle, fails to convince.

If it is true that defending a subjective view of morality implies an objective standard (the non-harm principle), the inclusion of objectivity seems inevitable. It is in this sense that one cannot do without objectivity.

Furthermore, some could argue that an individual could enhance in such a way as to become a unique type—particularly if one follows a view of authenticity in which one should be free to be whoever she wants. This position is also not satisfying because the notion of type itself is something shared, e.g. a shared humanity. One could argue that being the tallest woman in the world is in itself a unique type. However, this argument is not sound because she is in fact the tallest only in reference to other women.

Problems with an objective approach

Denial of pluralism

One problem with the objective view is the difficulty of reaching consensus in a liberal society in terms of what the objective goods are and what virtues or what types should be aimed for. Regarding a list of virtues, Harris, unlike Sandel, does not see humility as a virtue. Therefore, when taking an objective approach, one has to address the problem that some will want to impose their ideals on others, which could lead to forms of eugenics, as anyone who does not fit within an objective list or type would then be subject to some form of discrimination.

Here, however, we need to make a distinction between a legal and a moral position. Morally, Sandel is free, of course, to endorse humility as a virtue and to choose to live a humble life, even if someone else would not approve of it. This would become problematic only if Sandel tried to impose his moral view by passing a law.

Nonetheless, even in a pluralistic society, we recognize that some types of being are better than others in certain ways. This is why, for example, education is mandatory in most countries, despite the fact that this requirement may limit the autonomy of the individual or of the parents in certain respects. Even if a society decides not to make education mandatory, deciding not to intervene in someone else's life by imposing a certain ideal, education can still be viewed as better than the lack of education. This can be shown to be true by describing its benefits. When a new technology appears, nobody needs to be coerced to use it. If the technology is useful, an individual will choose to adopt it.

Type implies objectivity

Because I have argued that type-perfection is more plausible than property-perfection, I further contend here that a type-perfectionist approach also implies a commitment to objectivity. One could posit that this commitment to objectivity is not necessary because one could imagine what her best self could look like based on values shared only by herself, thereby defending that a type-perfectionist approach would be compatible with a subjective approach. However, as was mentioned earlier, this argument fails to grasp that the very nature of a type is that it is shared. We are thus left with the better option of an objective type-perfection. How, then, should we use this type-perfection?

Function of Perfection

How are perfectionist assumptions used in the debate? Here, I distinguish between two antagonistic ways that perfectionist assumptions are used: they function either as a constraint of human enhancement or as a guiding principle.[9] Those who employ perfectionist assumptions as a limitation of human enhancement state that our understanding of a good human life is shaped by the limitations humans currently possess (e.g. embodiment and finality). Because human enhancements are a threat to these limitations, which are seen as essential characteristics for a good life, they ought to be rejected. In other words, if certain limitations can be overcome by biotechnological means, humans would no longer be able to experience real flourishing. Their lives would become diminished, and their agency would be undermined. In this view, human enhancement is dangerous because it seeks to gain full control of human constitution and human lives by mastery over nature. In this context, limits are essential to a good life and should therefore not be tampered with. In the debate, this limitation view is used in two distinct ways. First, it is used against the view that human perfection means mastery over human nature (Kass, 2004; Sandel, 2007). Second, it is used against the post-human type (Agar, 2010), by defending a shared humanity.

---

9   For this section, see also *Human Dignity in Bioethics and Biolaw*, in which the authors investigate whether dignity should be an empowerment or a constraint (See Beyleveld and Brownsword, 2002).

The second way perfectionist assumptions are used is as empowerment to guide the use of human enhancement towards a certain ideal. For example, the subjective list suggested by Walker would help individuals to choose which human enhancement they should pursue and, by doing so, move towards a certain ideal. Objective goods would work in the same manner, as would the ideal of autonomy, in which individuals should pursue human enhancement to become masters of their own human nature (Agar, 2004; Harris, 2007; Brock, 2009).

It is unclear why perfectionist assumptions should only be used to limit or ban human enhancement. Indeed, one can hold some perfectionist assumptions and nevertheless accept that some forms of human enhancement would be helpful in trying to secure a fulfilling life. One can enhance within this particular type. For instance, while Fukuyama argues that human nature at its core is made of what he calls Factor X (Fukuyama, 2003), it is unclear whether human enhancements are necessarily a threat to Factor X, as they could also be used as a means to secure or improve Factor X.

Therefore, even for people who oppose human enhancement, the limitation view should not be applied to all enhancements—only to those that could be a threat to leading a flourishing life. Those who are against some enhancements might themselves be willing to endorse other enhancements that will help them lead a good human life. Like any other improvements, whether a human enhancement is interpreted as positive or not largely depends on one's view of the good human life.

However, arguing that perfectionist assumptions can be used only to guide human enhancement is also flawed. Indeed, an ideal can be used to limit, but it can also help direct decision-making and behavior. When someone is encouraged by her conception of the good human life to enhance in one direction, this will simultaneously lead her not to enhance in another direction. In this sense, limiting and guiding are two sides of the same coin. Perfectionist assumptions therefore should not be used only to constrain or ban human enhancement, but can also be used to guide and direct.

From human perfection to perfectionist notions

Depending on different social and historical contexts, we will not find consensus in a pluralistic society regarding what an ideal human being ought to be, act like, and look like. We can nonetheless discuss and agree upon some specific human characteristics (perfectionist assumptions) that appear to be essential for such an ideal human in contemporary society. These characteristics can then be used as reference points to assess the morality of human enhancements in addition to other normative tools, such as safety, justice and autonomy. With an end-state of human perfection in mind, –we can have a set of perfectionist assumptions that are essential to leading an ideal life, even if this ideal might continue to develop and evolve as societies change.

Perfectionist assumptions in the debate

Different perfectionist assumptions have been used in the debate to evaluate the morality of human enhancement. However, some of them are flawed because they do not fit the requirement of being an objective type-perfection that can both guide and limit human enhancement.

Michael Sandel

In *The Case against Perfection*, Sandel claims to be against perfection. He not only implies that proponents of enhancement seek perfection; he also argues that humans have some reasons not to pursue perfection. Sandel's real problem with enhancement, however, is not perfection itself, but the drive to attain mastery, as it is "mastery and dominion that fails to appreciate the gifted character of human powers and achievement" (Sandel, 2007, p. 83). For him, human enhancement is problematic because it does not recognize the giftedness of nature or of life. Sandel fears that enhancement will lead to a mastery and dominion of the world (Sandel, 2007) and argues that pursuing enhancement technologies is hubristic and could be a threat to the human virtues of humility, solidarity, and responsibility Sandel highly values.

To argue against human enhancement, Sandel first demonstrates that notions of justice, autonomy, and safety fail to build a case against the practice (Sandel, 2007). If these notions fail, additional normative tools are needed to evaluate enhancement. To do so, Sandel relies on a notion of type-perfection. For him, the ideal human should pursue virtues of humility, solidarity, responsibility and gratitude for what they are given because this is the "proper stance of human beings towards the given world" (Sandel, 2007, p. 9). This type-perfection should limit to what extent humans should be able to alter themselves. For Sandel, pursuing enhancement is morally dangerous because it moves away from the boundaries of type-perfection.

Sandel is therefore not against perfection *per se*. To the contrary, he defends his understanding of type-perfection and is against the enhancement of some traits that would enhance someone beyond the limits delineated by his understanding of human type-perfection. He might even be against anyone whose type-perfectionist views do not correspond with his own. Sandel is against:

1) type-perfectionist views that differ with him regarding the content of a particular view of type-perfection;
2) views that do not hold a particular type-perfection but have some ideas about how traits or properties ought to be perfected. For example, some individuals, while trying to achieve a 'perfect' speed or a 'perfect' intelligence for a certain task extend beyond the normative limits given by Sandel;
3) views that do not embrace type-perfection, but hold that autonomy, justice, and safety are enough to guide the enhancement project.

In this sense, *The Case against Perfection* is actually not a case against perfection. It is a case against the wrong reasons for seeking perfection, and it also proposes a Sandelian type-perfection.

Problems with Sandel's approach

It is not clear why maintaining that such virtues are essential to live a good human life would also require a rejection of the enhancement project. Human enhancement could also be a means for improving solidarity, humility, responsibility and gratitude, among other virtues someone wishes to pursue. In this light, Roache rightly asks: "Must conservatives endorse enhancement of conservative values?"[10] Moral enhancement, for instance, may be able to help people display virtues they hold as important. It has even been suggested that some types of enhancement may create more altruistic personalities (Persson and Savulescu, 2012). Sandel's type-perfection could therefore be used not only to limit human enhancement, but also to

---

10 See www.rebeccaroache.weebly.com/research.html. Accessed March 6, 2014.

encourage some types of human enhancement that would not be a threat to his own particular type-perfectionist view.

Some might disagree, however, with this particular view of type-perfection. Harris, for instance, does not think of humility as a virtue (Harris, 2007) and therefore positions himself against Sandel's perspective. Other concerns might include what would happen to individuals who desire to pursue other virtues that are not considered essential to a Sandelian type-perfection. Accordingly, this view has been interpreted to be a threat to pluralism. Harris rightly fears that someone else's conception of type-perfection would be imposed on him (Harris, 2007) because particular perfectionist assumptions would entail a particular set of required enhancements.

The President's Council on Bioethics and Leon Kass

Like Sandel, the former President's Council on Bioethics appears to be against perfection in *Beyond Therapy, Biotechnology and the Pursuit of Happiness* and accuses proponents of human enhancement of pursuing human perfection. As the President's Council writes: "Some celebrate the perfection-seeking direction in which biotechnology may be taking us. Indeed, some scientists and biotechnologists have not been shy about prophesying a better-than-currently human world to come, available with the aid of genetic engineering, nanotechnologies, and psychotropic drugs" (President's Council, 2003, p. 6).

For the Council, the problem with human enhancement is that it might pose a threat to human dignity (President's Council, 2003, p. 140–5). They also rely on a concept of type-perfection to argue against human enhancement. Like Sandel, the President's Council first argues that bioethical notions such as safety, efficacy, morality of the means, unfairness and inequality, coercion and constraint do not get to the core of the problem (President's Council, 2003, p. 138). The President's Council thus relies on an additional normative tool to evaluate enhancement in the concept of type-perfection, which for them is defined "not as a life lived with an ageless body or an untroubled soul, but rather a life live in rhythmed time, mindful of time's limits" (President's Council, 2003, p. 299).

Normative references are thus prescribed by a view of type-perfection, which consists of embodiment and finality. For the President's Council, accepting our embodiment and honoring the limit that this embodiment gives us (Lebacqz, 2011) is important: "What matters is that we produce the given result... in a human way as human beings, not simply as inputs who produce outputs" (President's Council, 2003, p. 153). Whereas Sandel's type-perfection is about human virtues, the type-perfection of the President's Council is about human biological limitations.

Serving as the chair of the President Council, Kass' views are closely related to those of the President's Council. At first, he attacks the notion of perfection, stating: "Let me begin by offering a toast to biomedical science and biotechnology: May they live and be well. And may our children and grandchildren continue to reap their ever tastier fruit – but without succumbing to their seductive promises of a perfect, better-than-human future, in which we shall all be as gods, ageless and blissful" (Kass, 2003, p. 9). Kass associates perfection with something 'better' than human, which for him has negative consequences. In this understanding, becoming perfect would have the terrible consequences of making someone post-human, which he sees as dehumanizing. His problem with perfection is that when someone aspires towards it, that individual might cease to be human (Kass, 2004).

Additionally, he accuses his opponents to be pursuing 'perfection,' writing:

> Dreams of human perfection – and the terrible consequences of pursuing it – are themes of Greek tragedy ... Until now these dreams have been pure fantasies, and those who pursued them came crashing down in disaster. But the stupendous successes over the past century in all areas of technology, and especially in medicine, have revived the ancient dreams of human perfection. (Kass, 2003, p. 14)

Kass is therefore strongly opposed to the view of type-perfection, which would include a "painless, suffering-free, and finally, immortal existence" (Kass, 2004, p. 132). It is not compatible with his own particular view of type-perfection. Like Sandel and the President's Council, after accusing his opponents of pursuing perfection, Kass also takes a normative stance by suggesting his own view of type-perfection. For him, type-perfection includes the notion that humans are mortal, finite, and embodied. These characteristics provide normative limits to circumscribe what ought to be done to humans.

For Kass, embodiment, finitude and mortality are seen as essential parts of type-perfection. If certain limitations on humans can be overcome by biotechnological means, it is argued that they will no longer able to experience real flourishing. Human lives would become meaningless, and their agency would be undermined.

Problems with the President's Council & Kass

Some problems arise with each of these positions. First, these assumptions about the importance of limitations must not necessarily apply to all human enhancements, but only those that could present a threat to these particular views of type-perfection. The reasoning that one would be against a cognitive enhancement because it might pose a threat to embodiment and finitude seems strange. How does taking a cognitive enhancer in fact threaten human embodiment and finitude? Is it because one needs to accept her limitations? If so, is it necessary to clarify the difference between learning with a cognitive enhancer and learning from a book? Nonetheless, this argument can be used to argue against more radical enhancements, such as mind-uploading, which would be a more obvious threat to the embodiment of humans. Proponents of this view need to clarify precisely which human enhancements they are against, instead of treating them all the same.

Second, this view accepts the status quo of the human species, which not everyone accepts. Just because humans currently are bound by some limitations, this does not mean that they ought to accept these limitations. This view falls into the 'is-ought' fallacy, which holds that because humans have such limitations, they ought to keep them. Although for some, death is an accepted limitation that gives meaning to life, this notion does not have to be accepted by all, as the third problem suggests.

This view offers a single standard of type-perfection. If someone does not agree with this particular type-perfection and chooses to enhance nonetheless, what would happen to her? It is the inherent paternalism that comes with these views that is most problematic because not everyone will agree with the limits given here.

Nick Bostrom

Although Bostrom has spoken directly against perfection – calling it a useless concept (Bostrom, 2011)[11] – he nonetheless relies on perfectionist notions. The distinction made between type and property-perfection

---

11 Bostrom writes: "I don't think that perfection is a useful concept. There is not necessarily one best form of human existence; perfection might be different for different people" (Bostrom, 2011).

helps clarify his view. For Bostrom, the notion of human type-perfection is not useful; nonetheless, he embraces a notion of property-perfection when he argues for super-health, super-emotion and super-cognition. For him, health, emotion, and cognition are three properties that should be perfected.

Bostrom wants to improve different functions to make them the best they can be, even if this results in transgressing the limits of human type-perfection. However, this still leads to a state of perfection, which could be referred to as post-human type-perfection. If we cannot agree on one singular view of type-perfection, we will certainly not be able to agree on a posthuman type-perfection.

Bostrom's response here moves from type-perfection to property-perfection. He recognizes that some objective goods, such as health, cognition, emotion (Bostrom, 2008), and rationality (Bostrom, 2008) are properties that could be perfected. He positions some objective goods that, if perfected, would enable humans to live better lives. He thus advocates for a particular *telos* for humankind, which will be shared by posthumans. He writes:

> It would be perfectly possible for a posthuman to realize a telos of rationality as well as a human being could. In fact, if what is good for us is to develop and exercise our rational nature, this implies that it would be good for us to become posthumans with appropriately enhanced cognitive capacities (and preferably with extended healthspan too, so that we may have more time to develop and enjoy these rational faculties). (Bostrom, 2008, p. 130)

Like Walker, Bostrom rejects type-perfection but embraces a notion of property-perfection. Whereas Bostrom outlines some objective goods, such as greater health, stronger emotions, and longer life span, Walker refuses to mention any, leaving individuals free to choose from themselves what states or abilities they wish to enhance (Walker, 2002).

Problems with Bostrom's approach

One of the problems with Bostrom's approach is that we either have to agree on some type of list of objective goods, or we have to allow everyone to pursue or enhance whatever subjective goods may seem right for them, or what properties they want to enhance. The former premise appears impossible because people do not uniformly agree about virtues, values or conceptions of a good life. Why should we accept Bostrom's list of objective goods, and not someone else's? The latter point is also problematic. What should we do, for example, if someone were to view 'submission' as a good with which she wants to 'enhance' her children or her slaves?

In addition to these problems, Bostrom's position has also "insufficiently engaged with the possibility that transhumanist technologies pose a threat to the meaningfulness of human social existence" (Edgar, 2009, p. 158). Indeed, if individuals were free to choose any properties they want to enhance to their limits, it would create beings unable to affiliate or relate with other beings. We will explore this problem in greater detail below.

Moreover, as was mentioned earlier, property-perfection fails to be a convincing position on its own because it inevitably falls back into type-perfection. In this case, it may be posthuman type-perfection, but it will nonetheless be a view of type-perfection that will need to be explained in more detail.

Nicholas Agar

As with the so-called bioconservatives previously cited, Agar also embraces a type-perfectionist approach to argue against some types of enhancement. He argues against what he calls radical enhancement and defends some types of shared humanity and species relativism (Agar, 2010). For him, some of the things that humans can experience now are now under threat by radical enhancement. Therefore, humans should remain human and not become posthumans, because becoming posthuman might alienate them from meaningful relationships, or make them treat other humans as slaves (Agar, 2010). Like Sandel, Kass and the President's Council, who use type-perfectionist notions to limit enhancement, Agar employs a similar argument, but only to limit what he calls radical enhancement.

Criticism of Agar

Agar's main difference with so-called bioconservatives has to do with the content of type-perfection. Instead of being against human enhancement, he is against any enhancement that would transform humans beyond their species, which would result in another type. For him, humans should not become posthumans because it might alienate them from meaningful relationships with other humans.

Agar rightly acknowledges that type-perfection does not have to reject all enhancements. As we will see later, Agar uses a similar argument to the one defended here. For him, some enhancements might be problematic because they may erode what Nussbaum refers to as the capability of affiliation: the ability for humans to relate to one another. However, Agar focuses only on one capability, i.e., the capability of affiliation. He does not provide, therefore, a fuller picture of a given type-perfection. The remainder of this work will address this shortcoming.

Chapter 5: Defending a particular view of perfection

Summary

After a critique of views of perfection that are problematic (Chapter 4), I take a positive approach (Chapter 5) in which I suggest some perfectionist elements based on the work of Martha Nussbaum's capabilities approach that can lend guidance to the practice of human enhancement. I suggest here that the central capabilities outlined in her work can be used to define the human aspect of human enhancement and thereby allow for a moral evaluation of enhancement interventions. These central capabilities can be maximized harmoniously in order to postulate what an ideal human would look like.

Introduction

Having taken a negative approach regarding what perfectionist assumptions we should rid ourselves of, I now suggest a positive approach in which I defend some perfectionist elements that are essential to the vision of an ideal human being, which in turn can be used to guide human enhancement. If an intervention moves closer to that ideal/reference point, this will count – prima facie – in favor of the respective enhancement intervention. This final chapter seeks to answer the following question: Which type-perfection should be used in the debate regarding the ethics of human enhancement? Nussbaum's capabilities approach offers one possible solution and has many advantages, as we will see.

Defining capabilities

The capabilities approach, famously developed by Sen and Nussbaum, is a theoretical framework that focuses on the moral importance of individuals achieving well-being, which is understood in terms of people's capabilities, that is, their real opportunities to do and be what they have reason to value (Nussbaum, 2011). As Wells explains: "This distinguishes it from more established approaches to ethical evaluation, such as utilitarianism or resourcism, which focus exclusively on subjective well-being or the availability of means to the good life, respectively" (Wells, 2012).

The capabilities approach was first articulated in economics research, after which it became extensively used in other fields, for example, by the United Nations Development Program, as a broader, deeper alternative to narrow economic metrics such as growth in gross domestic product (GDP) per capita. This 'human development' approach has been designed to remedy this problem (Nussbaum, 2011). As Nussbaum puts it, capabilities are the answer to the question: "What is this person able to do and to be?" (Nussbaum, 2011, p. 18).

Capabilities are often explained in comparison to functioning. People should have a set of opportunities (capabilities) that they are then free to exercise or not exercise (functioning) (Nussbaum, 2011). Functioning is therefore the realization of capabilities. "To use an example of Sen's, a person who is starving and a person who is fasting have the same type of functioning where nutrition is concerned, but they do not have the same capability because the person who fasts is able not to fast, and the starving person has no choice" (Nussbaum, 2011, p. 25).

Nussbaum outlines a list of ten central capabilities, which include life; bodily health; bodily integrity; senses, imagination, and thought; emotions; practical reason; affiliation; other species; play; and control over one's environment (Nussbaum, 2011). These capabilities are so fundamental for humans that without them, life would be seriously impoverished (Nussbaum, 2011). Without them, it would be impossible to think of type-perfection.

Reasons to look at the capabilities approach

The capabilities approach matches and fulfills the requirements for evaluating human enhancement that have been put forth in Chapter 4. Along with outlining the different reasons why it is helpful to look at the capabilities approach in the debate, I show here how the capabilities approach fulfills the requirement of an objective type-perfection framework that can be used to both guide and limit human enhancement.

Make it explicit

Some authors (e.g. Parens, 1995) have implicitly based their assumptions of evaluating human enhancement on capabilities. This deserves to be made explicit. Others have used the capabilities approach as a normative framework to assess human enhancement (Coeckelbergh, 2010). Others have focused on specific capabilities, arguing that human enhancement could be a threat to those capabilities, such as the capability of affiliation (Agar, 2010; Cabrera and Weckert, 2012). Still others have suggested that the capabilities approach should be used as framework to guide human enhancement (Hughes, 2011). Finally, some have argued that when it comes to evaluating human enhancement, a point of reference is needed, which can be found in what a good human life consists of and can be explained using capabilities (Baertschi, 2011). However, an explicit explanation of how the capabilities can be used is lacking in the existing literature.

Being human: objective type-perfection

For societies to function, there is a need to agree on what it means to be human or at least agree on a minimum common ground. The capabilities approach gives a possible explanation of some capabilities a human must possess in order to lead a flourishing life. The "central capabilities can be interpreted as an expression of what it is to be human" (Coeckelbergh 2010, p. 82). Therefore, the ideal human (towards which we are striving with human enhancement) should have those central capabilities. A set of enhanced central capabilities can thus be interpreted as an expression of an enhanced human.

Using the capabilities approach as an attempt to define what it means to be human is of great help in the debate over human enhancement because too often the 'human' part of the debate regarding human enhancement is neglected or does not receive due attention. The capabilities approach provides an outline of what central capabilities the ideal human should have, structured in a holistic manner. Below this level, humans would be far from ideal. However, there also exists the possibility to add, remove or alter capabilities, which could eventually lead someone to become something other than human. In this framework, the capabilities suggest a limit to *human* enhancement.

Ideal theory

Nussbaum's capabilities approach can be defined as an ideal theory because an objective list gives an end-point to aim towards. This view overlaps with arguments developed elsewhere, which show that an ideal approach is helpful in assessing the morality of human enhancement (Roduit et. al, 2015). Nussbaum's theory could be classified as an ideal *end-state* theory because she does endorse some end-state perfectionist thinking by giving a list of central capabilities. She does offer some elements that are essential for human type-perfection. Her theory is therefore an ideal theory because one needs a reference point to evaluate the better option between two choices. Capabilities help to measure the distance between one's current situation and the ideal level of capabilities. It is recognized here that to evaluate something as better, we can look at the distance between existing capabilities and ideal capabilities. Therefore, one way to measure human enhancement is to consider the distance between a maximum level of capabilities and one's current state.

End-state thinking

Such an ideal theory holds certain advantages. It provides a framework of an 'end-state' type of thinking and provides direction to human enhancement. This end-state, however, does not need to be fixed, unlike in the approaches of Sandel, Kass, and the former President's Council on Bioethics. With this approach, "capabilities – and hence what humans are – are not fixed but change together with our technological and social context" (Coeckelbergh, 2010, p. 86). This approach allows individuals to move towards their own view of the ideal human, within some limits, as the ideal human would have to have a basic set of capabilities to be included in the human type.

The preliminary list given by Nussbaum outlines initial 'end-points' that can be used to evaluate two alternative courses of action by looking at a reference point provided by the capabilities. If, for example, a human enhancement eroded a given capability, it would then be morally problematic. Such a human enhancement could be a threat to bodily integrity, the capability of affiliation, or another capability. In this context, the capabilities become essential, as they provide a reference point to help to evaluate an improvement.

Room for pluralism

Not having fixed end-states leaves room for public discussion about different end-states. Both "Sen and Nussbaum are keen to keep their formulations of capabilities at a general, vague level to keep their theory wide open to conceptual adjustment and applicable to many different societies and circumstances" (Coeckelbergh, 2010, p. 83). Such an open-ended list also has the advantage of being open to political deliberation (Coeckelbergh, 2010). This position allows for the option to adjust, change, or have a public discussion regarding each capability.

This approach therefore refocuses the discussion of human enhancement and encourages public discussion of which capabilities are worthy of pursuing (e.g., is being able to fly worth pursuing). It will then necessitate a further public debate "about which capabilities matter and who (how, when) is to decide this" (Oosterlaken, 2012, p. 5).

Holistic approach of the capabilities approach

This approach provides some preliminary tools to build an ethical framework to guide human enhancement (Coeckelbergh, 2010). "As an ethical framework ... the capabilities can be understood not only as minimal requirements or 'thresholds' for dignity or justice, as Nussbaum usually sees them, but rather as formulations of the ethical 'maximum': they can be interpreted as what the good life or human flourishing requires" (Coeckelbergh 2010, p. 84). Indeed, if the "central capabilities can be interpreted as an expression of what it is to be human" (Coeckelbergh, 2010, p. 82), one might be tempted to argue that the maximization of the same central capabilities could then be understood as an expression of what it is to be an enhanced human.

However, this approach is slightly misguided. Nussbaum's framework, which intends to guarantee the enjoyment of central capabilities to all human beings, does not imply that maximizing the central capabilities would necessarily be better: after all, more is not always better. To maximize one capability over another might even be counterproductive. For example, maximizing the capability of emotion over practical reason or affiliation might make some individuals unable to function in society. Additionally, if we emphasize only one dimension of capabilities, we fall back into some sort of property-perfection. A more productive approach here is to look at the central capabilities from a holistic perspective. I suggest here that it is preferable to maximize all capabilities together as a harmonious, holistic set.

Both constraining and guiding

Finally, the capabilities approach is helpful because it can be used as a theory of type-perfection that is – unlike other type-perfectionist views – both guiding and restricting. On the one hand, it is constraining because a given enhancement that would cause one to fall under the threshold of a central capability would be seen as morally problematic. At the same time, it is guiding because a given enhancement should move a human towards a holistic ideal human, as outlined by the central capabilities. It enables us, therefore, to evaluate the morality of a given human enhancement intervention by considering whether the capabilities approach would ban the given enhancement, or allow it. This framework does not take a general stance by banning or accepting all human enhancements; instead, it encourages a case-by-case analysis, as is recommended by some scholars (Mukerji and Nida-Rümelin, 2014), and it can be used as a lens or a matrix to analyze different types of human enhancement.

In this regard, one could look at how adding, removing, extending or decreasing a particular capability through an enhancing intervention would affect the central capabilities as a whole and see whether it would undermine, maximize or not change them at all. Answering how a particular enhancing intervention would impact the central capabilities, offers an additional normative tool to evaluate the morality of a given type of human enhancement.

Conclusions

The aims of Chapter 4 and 5 were twofold. First, I made explicit the underlying perfectionist assumptions found in the debate about human enhancement and argued that although the type-perfectionist view is correct, this view does not necessarily have to limit human enhancement. It can also guide it. This view avoids the problems of property-perfection, which ultimately collapses into type-perfection. It helps guide the enhancement project instead of uniformly limiting all human enhancements.

Having defended the concept of type-perfection, I then suggested a particular conception of type-perfection, drawing on Nussbaum's central capabilities approach to argue that the ideal human should at least have the ten central capabilities put forth by Nussbaum, and that a given human enhancement is morally acceptable if it maximizes the central capabilities in a holistic manner.

This sketch of what an ideal human should look like helps supplement the other normative tools provided in the debate to evaluate the morality of human enhancement. The picture of the ideal human becomes an extra reference point that can be used with other bioethical notions such as safety, justice, and autonomy.

The framework suggested here also encourages a continuous public discussion to justify and decide what the central capabilities are, and whether new capabilities should be included with the central ones. Once those central capabilities have been agreed upon, one is free to enhance within those limits to reach towards what one considers being one's ideal human self. This would therefore help maintain a certain unity within humankind, while also allowing for diversity. However, the approach does establish a safeguard to ensure that through human enhancement, humans develop towards ideal humans, and not subhumans. For future research, it would be beneficial to look into whether it is morally permissible for humans to use enhancing technologies to change their own type.

Concluding

Conclusion

The debate regarding the ethics of human enhancement is still in its infancy. Issues regarding what exactly constitutes enhancements and how to morally evaluate various human enhancements have not been entirely resolved. This work has provided some additional tools to better understand and approach the debate in the future. As new technologies become available, they will both challenge and likely alter our understanding of what it means to be human.

This research has added to this discussion by showing how the concept of human perfection can be used as an additional normative tool to evaluate the morality of human enhancement. This work has argued that refusing to look at human perfection in this debate would ensure serious shortcomings vis-à-vis the moral evaluation of human enhancement. Instead of trying to altogether avoid or only implicitly refer to perfectionist assumptions, I suggested embracing them and making them explicit. Doing so will better inform other participants in the debate and help them understand the reasons authors think of human enhancement as morally dubious or as morally mandatory.

The main goal of this work was to investigate how the notion of perfection should be used in the debate, if at all. After first showing that the concept of perfection is in itself unavoidable, I then presented some requirements that a conception of perfection should fulfill in order to be of assistance in this debate. I concluded by suggesting a possible conception of perfection, based on the work of Nussbaum and her capabilities approach, adding to other normative tools such as the notion of justice, safety, and autonomy. Therefore, encouraging different actors in this debate to be explicit regarding their own view of human perfection, instead of avoiding it.

In further research, it will be necessary to look at practical examples in more detail, and apply the theoretical framework to different enhancements. One will have to analyze each new enhancing technology on its own.

However, I have given a general framework that enables us to look at human enhancement from the broader perspective of the human, and not only from the specific capacities that are enhanced. The following literary example might help illustrate this point. Imagine for a moment that you were told to improve a text you have written by replacing each word of the text with another better word. While each word on its own seems to have improved, the overall text will have lost its meaning. In this sense, the text is no longer authentic. This example illustrates the importance of considering the text holistically. In an analogous way, it is crucial to look at the human holistically when attempting to assess the morality of human enhancement. Nussbaum's capabilities approach fosters this holistic view and encourages us to take into consideration all the central capabilities when one particular capacity is enhanced. It helps us to realize, for example, when analyzing the morality of life-extension that one needs to take into account how this enhancement would affect all the central capabilities, as a whole.

This example also demonstrates how using perfectionist notions to evaluate the morality of human enhancement differ from using other anthropological notions, such as human nature or human authenticity. While we may not know precisely what makes for a 'perfect text' (full-fledged perfection), as this depends on many considerations, such as the time period in which it was written, or the textual style, we can nevertheless draw upon some elements that are essential to a perfect text (perfectionist assumptions). These might include factors such as having no grammatical, spelling mistakes, or using words in the appropriate language. Similarly, the approach of this thesis does not defend a full-fledged view of human perfection, but instead provides some elements that are essential for human perfection, based on a harmonization of Nussbaum's central capabilities.

It would also be crucial to include an evaluation of the content of the text in itself, not just its grammatical or stylistic characteristics, in order to qualify it as a perfect text. If not, a text making an apology of evil, for example, could qualify as a perfect text because it does not have grammatical errors and is written in a consistent tone. Consequently, a perfect text must be considered perfect not only because of its structural properties, but also because of its content. Likewise, the central capabilities help us to say something about a particular conception of what would be a perfect human, without committing to a full-fledged view of human perfection.

The nature of a perfect text also differs from the nature of a text in the same way human perfection differs from human nature. A text with spelling and grammatical errors is still a text, but it is far from a perfect text, as it does not possess some of the elements essential to a perfect text (no spelling or grammatical errors). The idea of the perfect text gives some guidelines on how to improve the sample text. In the same way, a particular conception of human perfection can offer some guidelines and goals for humans to improve towards.

Furthermore, a perfect text differs from an authentic text. A text might be authentic and yet be far from perfect due to some mistakes. Therefore, it can still be improved according to the standard outlined by the elements that a perfect text requires. Here, we see how the approach of this thesis differs therefore from other anthropological arguments in the debate.

Comparing the framework developed in this work with the bioconservative views can also help to clarify the main argument. For bioconservatives, such as Kass, Fukuyama or Sandel, human enhancement is morally unacceptable, since it is considered a threat to what it means to be human or to perfectionist notions of what it means to lead a good human life. However, a few problems are found in the bioconservative approach. Embracing some perfectionist notions or a particular view of human perfection does not necessarily imply that one should be against human enhancement. On the contrary, one might want to use some enhancement in order to improve one's condition, in order to be better aligned with one's particular view of human perfection. There are no apparent reasons why the conception human perfection could not change over time, as what it means to be human can also evolve over time.

Contrary to the bioconservative perspectives, the view defended in this thesis represents the ideal human we aim to become while using enhancement. This conception of human perfection encourages human enhancement in some circumstances, while prohibits it in others, depending on whether the given enhancements help the human being come closer to a given particular conception of human perfection. This conception of human perfection – based on the central capabilities – could change over time, as they are fluid definitions, not fixed ones.

This work, however, does not address the question of whether it is morally acceptable to enhance from one type to another. My sense here is that an approach based on Nussbaum's central capabilities approach could also be applied to other discussions that involve enhancement, as one could debate, for example, what central capabilities a posthuman should have. The debate of human enhancement could also become a debate regarding the enhancement of different persons or of different animals. We will need to delineate the central capabilities that posthumans or animals should have in order to use them as an additional normative tool to analyze, in this case, the morality of posthuman or animal enhancement (as was the case here with the discussion of human enhancement). New questions as to whether animals should be enhanced, or whether humans have a moral responsibility to enhance animals, will likely arise. But towards what end shall we enhance other animals? Should we make them other than animals? What about if some humans on the other hand desire to enhance in order to become like animals? Here questions regarding enhancement from one type to another (or from one species to another) will need to be addressed.

This type of questioning clearly shows that while the debate regarding human enhancement has been ongoing, it is only in an early stage. Technological progress promises to make us better. But, as the line between human enhancement and dis-enhancement, and between human, posthuman, or sub-human become more and more blurry, it will become more and more difficult to understand what 'better' actually means. It is therefore crucial to think through these issues, to think about this world we enter into, which will be far from perfect in spite of enhancement.

Finally, the conclusion of this work also questions the scope and limits of bioethics in general. Further research will be needed to investigate whether the findings and conclusions from the specific discussion in regards to human enhancement and perfection can be generalized to other bioethical debates. Should bioethicists become more receptive to discussions about ends and ideals instead of confining themselves to established bioethical principles? What can we learn from the debate about enhancement and perfection, in which the agnostic stance about perfection is untenable? Do the arguments in favor of this claim apply to other biomedical topics, too? And what would the bioethical debate look like if philosophers were to engage in public discussions and take into account the social dimensions of certain practices? These and other questions will need to be further developed and made relevant to the general discourse regarding the scope and limits of bioethics.[12]

---

12 See also Evans, 2012.

Bibliography

Agar, N. (2004). *Liberal Eugenics: in defence of human enhancement.* Blackwell Publishing.

Agar, N. (2010). *Humanity's End: Why We Should Reject Radical Enhancement (Life and Mind: Philosophical Issues in Biology and Psychology).* A Bradford Book.

Agar, N. (2014). *Truly Human Enhancement: A Philosophical Defense of Limits.* MIT Press.

Allhoff, F., Lin, P., Moor, J., & Weckert, J. (2009). Ethics of human enhancement: 25 questions & answers. *Ethics, Law, and Technology*, 1–50.

Allhoff, F., Lin, P., & Steinberg, J. (2011). Ethics of human enhancement: an executive summary. *Science and Engineering Ethics*, 17(2), 201–12.

Annas, G. J., Andrews, L. B., & Isasi, R. M. (2002). Protecting the Endangered Human : Toward an International Treaty Prohibiting Cloning and Inheritable Alterations. *American Journal of Law & Medicine*, 28(2&3), 151–178.

Baertschi, B. (2009). Inented changes are not always good, and unintended changes are not always bad–why? *The American Journal of Bioethics*, 9(5), 39–40.

Baertschi, B. (2011). L'Humanité se dit de multiples manières. *Journal International de Bioéthique*, 22(3–4), 67–76.

Baumann F. (2010). Humanism and transhumanism. *The New Atlantis*, 29: 68–84.

Beyleveld, D., & Brownsword, R. (2002). *Human Dignity in Bioethics and Biolaw.* Oxford University Press.

Bonte P. (2013). Dignified Doping: Truly Unthinkable? An Existentialist Critique of 'Talentocracy' in Sports. In J. Tolleneer, S. Sterckx, & P. Bonte (Eds.), *Athletic Enhancement, Human Nature and Ethics* (Vol. 52, pp. 59–86). Springer.

Bostrom N. (2005). In defense of posthuman dignity. *Bioethics*, 19(3).

Bostrom, N. (2008). Why I Want to be a Posthuman When I Grow Up. In B. Gordijn & R. Chadwick (Eds.), *Medical Enhancement and Posthumanity* (pp. 107–137). Springer.

Bostrom, N. (2011). Genetic Enhancement and the Future of Humanity. The European. Retrieved from http://theeuropean-magazine.com/282-bostrom-nick/283-perfection-is-not-a-useful-concept.

Brock, D. W. (2009). Is Selection of Children Wrong? In J. Savulescu & N. Bostrom (Eds.), *Human Enhancement*. Oxford University Press.

Brownsword, R. (2012). Five Principles for the Regulation of Human Enhancement. *Asian Bioethics Review*, (June), 344–354.

Brownsword, R. (2013). A Simple Regulatory Principle for Performance-Enhancing Technologies: Too Good to Be True? In J. Tolleneer, S. Sterckx, & P. Bonte (Eds.), *Athletic Enhancement, Human Nature and Ethics* (Vol. 52, pp. 291–310). Springer.

Buchanan, A. (2011). *Better than Human: The Promise and Perils of Enhancing Ourselves (Philosophy in Action)* (p. 208). Oxford University Press.

Buchanan, A., Brock, D. W., Daniels, N., & Wikler, D. (2001). *From Chance to Choice: Genetics and Justice*. Cambridge University Press.

Buchanan, A. (2011). *Beyond Humanity?: The Ethics of Biomedical Enhancement (Uehiro Series in Practical Ethics)*. Oxford University Press.

Cabrera, L., & Weckert, J. (2012). Human Enhancement and Communication: On Meaning and Shared Understanding. *Science and Engineering Ethics*. doi:10.1007/s11948-012-9395-2.

Caplan A. (2009). Good, better, or best? In Bostrom N. Savulescu J. (Eds.), Human enhancement (pp. 199–210). Oxford University Press.

Caplan A., & Elliott, C. (2004). Is it ethical to use enhancement technologies to make us better than well? *PLoS Med*, 1: e52.

Chadwick, R. (2008). Therapy, enhancement and improvement. In B. Gordijn & R. Chadwick (Eds.), *Medical Enhancement and Posthumanity* (pp. 25–37). Springer.

Chadwick, R. (2011). Enhancements: improvements for whom? *Bioethics*, 25(4).

Coeckelbergh, M. (2010). Human development or human enhancement? A methodological reflection on capabilities and the evaluation of information technologies. *Ethics and Information Technology*, 13(2), 81–92.

Cohen, E. (2003). Bioethics in Wartime. *The New Atlantis*, 23–33.

Daniels, N. (2000). Normal Functioning and the Treatment-Enhancement Distinction. *Cambridge Quarterly of Healthcare Ethics*, 9(03), 309–322.

DeGrazia, D. (2000). Prozac, Enhancement and Self-Creation. *The Hastings Center Report, 30*(2), 34–40.

DeGrazia, D. (2005). Enhancement technologies and human identity. *The Journal of Medicine and Philosophy, 30*(3), 261–83.

Edgar, A. (2009). The hermeneutic challenge of genetic engineering: Habermas and the transhumanists. *Medicine, Health Care, and Philosophy, 12*(2), 157–67.

Earp, B. D., Sandberg, A., Kahane, G., & Savulescu, J. (2014). When is diminishment a form of enhancement? Rethinking the enhancement debate in biomedical ethics. *Frontiers in Systems Neuroscience, 8*(12).

Erler, A. (2012). One Man's Authenticity is Another Man's Betrayal: A Reply to Levy. *Journal of Applied Philosophy*, doi:10.1111/j.1468-5930.2012.00562.x.

Evans, J. H. (2012). *The History and Future of Bioethics*. Oxford University Press.

Fukuyama, F. (2003). *Our Posthuman Future: Consequences of the Biotechnology Revolution*. Picador.

Foss, M. (1946). *The Idea of perfection in the western world*. Princeton University Press.

Grunwald, A. (2009). *Human Enhancment – What Does "Enhancement" Mean Here?* Retrieved from http://www.ea-aw.org/fileadmin/downloads/Newsletter/NL_88_042009.pdf.

Habermas, J. (2003). *The Future of Human Nature*. Polity.

Hanson, M. J. (1999). Indulging anxiety: human enhancement from a Protestant perspective. *Christian Bioethics, 5*(2), 121–38.

Harris, J. (2007). *Enhancing Evolution: The Ethical Case for Making Better People*. Princeton University Press.

Heilinger, J.-C. (2010). *Anthropologie und Ethik des Enhancements (Humanprojekt/Interdisziplinare Anthropologie) (German Edition)*. De Gruyter.

Heilinger, J.-C. (2014). Anthropological Arguments in the Ethical Debate. *Humana.Mente – Journal of Philosophical Studies*, (26), 95–116.

Hughes, J. (2011). After Happiness, Cyborg Virtue. *Free Inquiry, 32*(1), 1–7.

Huxley, A. (2006). *Brave New World* (Reprint ed.). Harper Perennial Modern Classics.

Hyde, M. J. (2010). *Perfection, coming to terms with being human*. Baylor University Press.

Juengst, E. (1998). The meaning of enhancement. In E. Parens (Ed.), *Enhancing Human Traits: Ethical and Social Implications* (pp. 29–47). Georgetown University Press.

Kass, L. R. (2004). *Life Liberty & the Defense of Dignity: The Challenge for Bioethics*. Encounter Books.

Kass, L. R. (2003). Ageless Bodies, Happy Souls : Biotechnology and the Pursuit of Perfection. *The New Atlantis*, 9–28.

Keenan, J. F. (1999). "Whose Perfection is it Anyway?": A Virtuous Consideration of Enhancement 1. *Christian Bioethics*, 5(2), 104–120.

Lebacqz, K. (2011). Dignity and Enhancement in the Holy City. In Cole-Turner R. (Ed.), *Transhumanism and Transcendence, Christian Hope in an Age of Technological Enhancement*. Georgetown University Press.

Levy, N. (2011). Enhancing Authenticity. *Journal of Applied Philosophy*, 28(3), 308–318.

Mahootian, F. (2012). Ideals of Human Perfection: A Comparison of Sufism and Transhumanism. In Tirosh-Samuelson, H., & Mossman, K. L. (Eds.), *Building Better Humans?: Refocusing the Debate on Transhumanism*. Peter Lang Pub Inc.

McKean, E. (2005). The new Oxford American dictionary. Oxford University Press, Inc.

Mckenny, G. P. (1999). Enhancements and the Quest for Perfection. *Christian Bioethics*, 5(2), 99–103.

McKibben, B. (2004). *Enough: Staying Human in an Engineered Age*. St. Martin's Griffin.

Mehlman, M. J. (2009). *The Price of Perfection: Individualism and Society in the Era of Biomedical Enhancement*. The Johns Hopkins University Press.

Menuz, V. (forthcoming). Why do we wish to be enhanced? In *Inquiring into human enhancement: beyond disciplinary and national boundaries*. S. Bateman, J. Gayon, S. Allouche, J. Goffette and M. Marzano (Eds). Palgrave McMillan.

Menuz, V., Hurlimann, T., & Godard, B. (2011). Is Human Enhancement also a Personal Matter? *Science and Engineering Ethics*. doi:10.1007/s11948-011-9294-y.

Mitchell, C. Ben. (2009). On Human Bioenhancements. *Ethics & Medicine: An International Journal of Bioethics, Volume 25*(3).

Mukerji, N., & Nida-Rümelin, J. (2014). Towards a Moderate Stance on Human Enhancement. *Humana.Mente – Journal of Philosophical Studies, 26*.

Naam, R. (2005). *More Than Human: Embracing the Promise of Biological Enhancement*. Broadway Books.

Nussbaum, M. C. (2011). *Creating Capabilities: The Human Development Approach*. Belknap Press of Harvard University Press.

Oosterlaken, I. (2012). The Capability Approach, Technology and Design: Taking Stock and Looking Ahead. In I. Oosterlaken & van den Hoven (Eds.), *The Capability Approach, Technology and Design*. Springer.

Passmore J. (2000). *The perfectibility of man*. 3rd edn. Indianapolis: Liberty Fund.

Parens, E. (1995). The Goodness of Fragility : On the Prospect of Genetic Technologies Aimed at the Enhancement of Human Capacities. *Kennedy Institute of Ethics Journal, 5*(2), 141–153.

Parens, E. (2005). Authenticity and Ambivalence: Toward Understanding the Enhancement Debate. *The Hastings Center Report, 35*(3), 34–41.

Pence, G. E. (2012). *How to Build a Better Human: An Ethical Blueprint*. Rowman & Littlefield Publishers.

Persson, I., & Savulescu, J. (2012). *Unfit for the Future: The Need for Moral Enhancement*. Oxford University Press.

President's Council on Bioethics. (2003). *Beyond therapy: Biotechnology and the pursuit of happiness*. Dana Press.

Rawls, J. (1999). *A Theory of Justice* (Revised ed.). Belknap Press.

Resnick, D. B. (2000). The Moral Significance of the Therapy-Enhancement Distinction in Human Genetics. *Cambridge Quarterly of Healthcare Ethics, 9*, 365–377.

Robeyns, I. (2012). Are transcendental theories of justice redundant? *Journal of Economic Methodology, 19*(2), 159–163.

Roduit, J. A. R., Baumann, H., & Heilinger, J.-C. (2013). Human enhancement and perfection. *Journal of Medical Ethics, 39*(10), 647–50.

Roduit, J. A. R., Menuz, V., & Baumann, H. (2014). Human enhancement: Living up to the ideal human. In S. J. Thompson (Ed.), *Global issues and ethical considerations in human enhancement technologies* (pp. 54–66). IGI Global.

Roduit, J. A. R., Baumann, H., & Heilinger, J-C. (2015). Evaluating human enhancements: The importance of ideals. *Monash Bioethics Review*, 32(3), 205–216.

Roduit, J. A. R., Heilinger, J.-C. and Baumann, H. (2015). Ideas of Perfection and the Ethics of Human Enhancement. *Bioethics*, 29: 622–630. doi:10.1111/bioe.12192.

Rubin, C. T. (2004). Man or Machine? *The New Atlantis*, 31–37.

Sandel, M. J. (2007). *The Case against Perfection: Ethics in the Age of Genetic Engineering* (p. 176). Belknap Press of Harvard University Press.

Savulescu, J. (2006). Justice, fairness, and enhancement. *Annals of The New York Academy of Sciences*, 1093, 321–338.

Savulescu, J. (2007). In defence of Procreative Beneficence. *Journal of Medical Ethics*, 33(5), 284–8.

Savulescu, J. (2009). Gentic Enhancement. In Kuhse H. and Singer P. (Eds), *A Companion to Bioethics, 2nd Edition* (pp. 216–234). Wiley-Blackwell.

Savulescu, J., & Bostrom, N. (2011). *Human Enhancement*. Oxford University Press.

Schaefer, G. O., Kahane, G., & Savulescu, J. (2013). Autonomy and Enhancement. *Neuroethics*. doi:10.1007/s12152-013-9189-5.

Schermer, M. (2008). On the argument that enhancement is "cheating". *Journal of Medical Ethics*, 34(2), 85–8.

Schramme, T. (2002). Natürlichkeit als wert. *Analyse & Kritik*, 2, 249–271.

Sen, A. (2006). What Do We Want from a Theory of Justice? *The Journal of Philosophy*, 103(5), 215–238.

Sen, A. (2009). *The Idea of Justice*. Allen Lane.

Shuman, J. (1999). Desperately seeking perfection: Christian discipleship and medical genetics. *Christian Bioethics*, 5(2), 139–53.

Simmons, A. J. (2010). Ideal and Nonideal Theory. *Philosophy & Public Affairs*, 38(1), 5–36.

Sparrow, R. (2011). Liberalism and eugenics. *Australasian Journal of Philosophy*, 89(3), 499–517.

Valentini, L. (2011). A Paradigm Shift in Theorizing About Justice? A Critique of Sen. *Economics and Philosophy*, 27(03), 297–315.

Valentini, L. (2012). Ideal vs. Non-ideal Theory: A Conceptual Map. *Philosophy Compass, 7*(9), 654–664.

Vita-More, N. (2013). The Future of Humanity. The European. Retrieved from http://www.theeuropean-magazine.com/natasha-vita-more--2/6564-the-future-of-humanity.

Walker, M. (2002). What is Transhumanism? Why is a Transhumanist? *http://www.transhumanism.org/*. Retrieved May 22, 2012, from http://www.transhumanism.org/index.php/th/more/298/.

Walker, M. (2013). *Happy-People-Pills For All (Blackwell Public Philosophy Series)*. Wiley-Blackwell.

Wall, S. (2008). Perfectionsism in moral and political philosophy. In E. N. Zalta (Ed.), *The Stanford Encyclopedia of Philosophy*. Retrieved from http://plato.stanford.edu/archives/fall2008/entries/perfectionism-moral.

Wells, T. (2012). Sen's Capability Approach. *Internet Encyclopedia of Philosophy*. Retrieved from http://www.iep.utm.edu/sen-cap/.

Zylinska, J. (2010). Playing God, Playing Adam: The Politics and Ethics of Enhancement. *Journal of Bioethical Inquiry, 7*(2), 149–161.

Index

A
Agar, Nicholas 13, 14, 102
Anthropology
  Dehumanization,
    Dehumanized 46, 52
  Human authenticity 26, 35,
    46–48, 87, 116
  Human being, fixed view vs.
    non-fixed view 19, 25, 27,
    28, 31, 32, 43, 44, 46, 50,
    53, 59, 61, 62, 64, 76, 78,
    80, 89, 95, 96, 99, 101,
    105, 109, 117
  Human dignity 24, 26, 46, 47,
    49, 52, 87, 92, 94, 98, 119
  Human nature 24, 26, 35, 37,
    43, 46, 59, 60, 62, 64–66,
    87, 90, 92, 94, 95, 116,
    119, 120, 121
  Human perfection 13, 19,
    26–28, 31, 33, 34, 57, 58,
    60–67, 72, 78, 79, 82, 83,
    87, 88, 94–96, 98, 99,
    115–117, 122
  Ideal Human 17, 26, 27,
    29, 33, 36, 39, 47, 48, 50,
    51, 53, 58, 59, 66, 71, 76,
    87–89, 95, 96, 105, 107,
    108, 110, 111, 117, 123
  Ideal Self 26, 46, 51, 52
Autonomy 19, 25, 26, 29, 30,
  31, 34, 35, 37, 39, 42, 45, 46,
  48, 50–52, 57, 58, 60, 62, 65,
  66, 80, 87, 88, 93, 95–97, 111,
  115, 124
  Decisional autonomy 45
  Functional autonomy 45

B
Bioconservatives 24, 28, 29, 34,
  35, 57–60, 62–67, 102, 117
Bioliberals 24, 28, 30, 34, 35,
  51, 57–60, 62–67
Bostrom, Nick 28, 29, 46,
  100–102, 119, 120, 124
Brave New World 25, 42, 121
Brock, Dan 29, 63, 80, 90, 95,
  120
Buchanan, Allen 29, 30, 40–42,
  44–46, 58, 61, 63, 72, 73, 87,
  92, 120

C
Capabilities approach 19, 31, 32,
  34, 36, 88, 92, 105–107, 109,
  110, 115–117
  And human functioning 23,
    40, 48, 63
  Adding capabilities 23, 33, 53,
    60, 78, 110, 115
  Central capabilities 19, 32,
    33, 35, 36, 88, 105–111,
    116–118
  Decreasing capabilities 36,
    110
  Increasing capabilities 71
  Removing capabilities 33, 110
Caplan, Arthur 42, 58, 63, 64,
  120
Chadwick, Ruth 40, 41, 72, 73,
  119, 120

D
Daniels, Norman 41, 72, 120
DeGrazia, David 46, 87, 121

127

Disease 23, 24, 40, 41, 81
Diversity 30, 32, 44, 82, 111

E
Elliott, Carl 63, 120
Enhancement
   Animal Enhancement 36, 37, 118
   Cognitive Enhancement 37, 100
   Cosmetic surgery 40
   Enhancing children 45
   Height enhancement 48, 78
   Human (see human enhancement)
   Posthuman Enhancement 36, 118
   Vs Therapy 40, 41, 72
Erler, Alexandre 15, 46, 87, 121
Eugenics 25, 43, 91, 93, 119, 124

F
Fukuyama, Francis 37, 42, 43, 46, 58, 87, 90, 95, 117, 121
   Factor X 95

G
God 43
   Playing God argument 125

H
Habermas, Jürgen 42, 58, 87, 121
Harris, John 30, 41, 44, 45, 48, 51, 63–65, 72, 73, 79–82, 90, 92, 93, 95, 98, 121
Health 16, 26, 32, 33, 35, 40, 101, 106, 120, 121, 123
Hubris 43, 52, 60, 96
Human enhancement 13–17, 19, 23–29, 31–36, 39–53, 57, 58, 60, 63–65, 71–81, 83, 87–91, 94–98, 100, 102, 105–111, 115–124
   Backward-looking view 24, 30, 34, 48, 49, 71–76, 79, 81, 83
   Dis-enhancement 19, 118
   Forward-looking view 24, 30, 34, 48, 49, 71, 72, 74, 76, 83
   Goals of – 64, 67, 68
   Personal optimum state 47, 91
   Qualitative view 9, 23, 34, 39–42, 47, 53, 72, 73, 77, 78
   Quantitative view 9, 23, 40–42, 72, 77, 78
Huxley, Aldous 25, 26, 42, 121

I
Ideal theory 71, 75–77, 79, 80, 82, 107, 108, 125
   vs. Non-ideal theory 71, 74–76, 82, 125

J
Juengst, Eric 72, 122
Justice 19, 25–27, 29–31, 34, 35, 37, 39, 42, 44, 45, 48, 52, 58, 60, 62, 66, 71, 74–77, 80, 82, 87, 88, 96, 97, 109, 111, 115, 120, 123, 124
   Fairness 26, 92, 124

K
Kass, Leon 24, 28, 37, 46, 47, 57–60, 62, 87, 90, 92, 94, 98–100, 102, 108, 117, 122
Keenan, James 30, 73, 74, 89, 122

L
Levy, Neil 46, 87, 121, 122

M
Mastery 59, 60, 62, 65, 66, 78, 92, 94, 96
  over one's life 35
McKibben, Bill 42–44, 58, 87, 122
Menuz, Vincent 15, 17, 40, 41, 45, 47, 73, 91, 122, 123

N
NBIC, *Nano-, Bio-, Info-Cogno-* 91
Neutrality 11, 28, 76, 78, 79
Non-ideal theory 71, 74–76, 82, 125
Nozick, Robert 26
Nussbaum, Martha 19, 31, 32, 34, 50, 88, 92, 103, 105–110, 115–117, 123
  Capabilities approach 19, 31, 32, 34, 36, 88, 92, 105–107, 109, 110, 115–117

P
Parens, Erik 46, 87, 90, 106, 122, 123
Passmore, John 58, 123
Paternalism 100
Perfection
  Concept of perfection 13, 14, 19, 27–31, 34–36, 57–59, 63–66, 115
  Conception of perfection 19, 27, 28, 31, 32, 34, 36, 57–59, 115
  Function of Perfection 94
  Good life, Good human life 19, 26, 29, 30, 39, 47–51, 53, 57–59 61, 62, 67, 74, 82, 83, 88, 91, 92, 94, 95, 97, 101, 105, 107, 109, 117
  Perfectionism 30
  Perfectionist notions 19, 28–30, 34, 36, 39, 47, 48, 83, 90, 95, 100, 102, 116, 117
  Property-perfection 31, 49, 89–91, 94, 100–102, 109, 110
  Sources of Perfection 91, 92, 93
  Pursuit of Perfection 28, 29, 57, 63, 66, 122
  Perfection as mastery over human nature 59, 62, 66
  Type-perfection 31, 32, 36, 49, 50, 89, 91, 94, 96–103, 105–108, 110
Pluralism 19, 31, 81, 93, 98, 108
Posthuman 16, 24, 25, 36, 37, 60, 62, 88, 90, 101, 102, 117, 118, 120, 121
Posthuman dignity 46, 119
President's Council on Bioethics 57, 98, 108, 123
Public discussion 32, 50, 51, 53, 108, 109, 111, 118
Public debate 32, 50, 109

R
Rawls, John 27, 76, 123

S
Safety 19, 25, 26, 29, 31, 34, 35, 37, 39, 42–44, 48, 52, 58, 60, 62, 66, 80, 87, 88, 96–98, 111, 115
  Non-harm principle 92
Sandel, Michael 11, 13, 14, 28, 37, 44, 46, 57–61, 63, 73, 82, 89, 90, 92–94, 96–99, 102, 108, 117, 124

Savulescu, Julian 15, 16, 25, 40, 73, 92, 97, 120, 121, 123, 124
Sen, Amartya 50, 74, 75, 79, 80, 81, 92, 105, 106, 108, 124, 125
Soma 25, 26, 42
Sparrow, Robert 44, 124

T
Therapy 23, 40, 41, 42, 72, 98, 120, 123
Transhumanism 29, 119, 122, 125

Transhumanist 24, 25, 36, 57, 102, 121, 125

V
Vita-More, Natasha 29, 125

W
Walker, Mark 30, 31, 48, 49, 65, 73, 74, 77, 89–91, 95, 101, 125

www.ingramcontent.com/pod-product-compliance
Ingram Content Group UK Ltd.
Pitfield, Milton Keynes, MK11 3LW, UK
UKHW041913140426
5217IPUK00002B/23